MONOGRAPHS ON
PHYSICAL BIOCHEMISTRY

GENERAL EDITORS

W. HARRINGTON   A. R. PEACOCKE

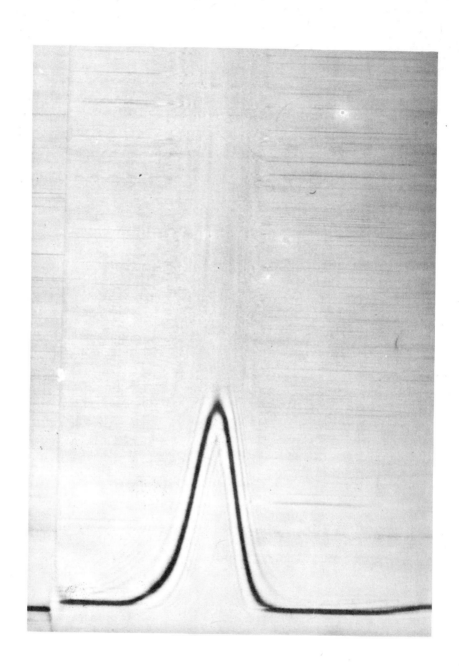

OPTICAL METHODS IN ULTRACENTRIFUGATION, ELECTROPHORESIS, AND DIFFUSION

WITH A GUIDE TO THE INTERPRETATION OF RECORDS

BY

PETER H. LLOYD

CLARENDON PRESS · OXFORD

1974

Oxford University Press, Ely House, London W.1

GLASGOW NEW YORK TORONTO MELBOURNE WELLINGTON
CAPE TOWN IBADAN NAIROBI DAR ES SALAAM LUSAKA ADDIS ABABA
DELHI BOMBAY CALCUTTA MADRAS KARACHI LAHORE DACCA
KUALA LUMPUR SINGAPORE HONG KONG TOKYO

ISBN 0 19 854605 X

© OXFORD UNIVERSITY PRESS 1974

PHOTOTYPESET BY OLIVER BURRIDGE FILMSETTING LTD,
CRAWLEY, SUSSEX

PRINTED IN GREAT BRITAIN
BY FLETCHER & SON LTD, NORWICH

PREFACE

ALTHOUGH the technique of free-boundary electrophoresis has now been largely abandoned in favour of electrophoresis on various support media, ultracentrifugation remains one of the most powerful methods available for the study of the physical properties of macromolecules. Modern advances in our understanding of the theory of these methods are requiring greater precision of experimental results. This, in turn, means that the individual scientist must have a greater understanding of the ultracentrifuge itself, and of the optical systems that it uses to display the distributions of concentration within the rotating cell. This book has been written to help the chemist or biochemist, with little knowledge of optics, to gain an appreciation of the principles involved, and to help him obtain the best possible results from his experiments. The first chapter is an elementary introduction to those optical properties of solutions that are used in the optical systems, after which the three basic optical systems are described in detail. Chapter 5 describes methods for the alignment of the optical systems, and Chapter 6 contains a fairly detailed account of the ways in which measurements must be made of the photographs in order to obtain the information that the system is being used to determine. Finally, Chapter 7 contains a comparison of the systems in which I have tried to review their relative merits. Because it is so often difficult to obtain truly objective criteria for comparison, this review contains many of my personal opinions which might not necessarily be shared by other workers in this field.

Throughout the text it is frequently necessary to distinguish the three dimensions in space. The convention that I have used throughout the book is set out at the end of the first chapter and illustrated in Fig. 1.2; thereafter the axes are frequently referred to without further definition. The fact that vertical coordinates in the cell are shown as horizontal coordinates in the diagrams is confusing, but is an almost universal convention. I have tried not to use such terms as 'vertical' or 'horizontal' unless the meaning is quite unambiguous. The photographs of the optical patterns are reproduced as they would normally appear on the photographic plates—i.e. they are negatives.

There are, of course, several different commercial forms of these optical systems available, fitted to different instruments. I have tried to keep the discussion as general as possible and not to indicate my personal preferences. However, I have felt it desirable to point out the occasional differences between a particular common commercial form of a system and the form as I have described it. There are two major manufacturers of these systems with which I am familiar: Beckman Instruments Inc., Palo Alto, California, U.S.A. and Measuring and Scientific Equipment Ltd., Crawley, Sussex, England. These are referred to simply as Beckman and M.S.E. respectively.

Finally I must express my appreciation to all my colleagues with whom discussions have helped me to clarify, in my own mind, the details described in this book. I am particularly indebted to two friendly physicists, who have patiently tried to enlighten me on some of those aspects of optical theory which, although elementary to them, were not so to a biochemist. I am also indebted to Beckman Instruments Inc. for permission to reproduce Fig. 1.1 and to M.S.E. Ltd. for supplying me with Figs. 2.4 and 2.5. I should like to thank the Medical Research Council for financial support during the writing of this book.

Oxford   PETER H. LLOYD
July 1972

CONTENTS

LIST OF SYMBOLS

a the distance through a solution in the direction of the optical axis.

A the area under a schlieren curve.

c concentration.

D the deflection of the pen of a densitometer.

E the extinction of a solution.

I the intensity of light transmitted by a solution.

I_0 the intensity of light incident on a solution.

J the refractive increment of a solution in terms of interference fringes and defined by eqn (6.7), p. 73.

L the optical lever in the schlieren optical system (see p. 69).

m the magnification factor relating vertical dimensions in the cell to horizontal measurements on the photograph; i.e. $X = mx$.

m' the magnification in the z direction between the schlieren analyzer and the photographic plate in schlieren optics.

n refractive index.

n' the refractive index of the external medium (usually air).

n_0 the refractive index of the reference solution.

r a positional coordinate for focusing the camera lens and defined on p. 88

ΔS the difference in optical path length for the two beams of an interferometer.

t time.

x the linear coordinate in the cell in the direction of applied forces and referred to as vertical.

X horizontal measurements on the photographs.

y the direction of the optical axis.

z the direction in the cell at right-angles to x and y, and referred to as horizontal.

Z vertical measurements on the photographs.

α the angle of deflection of a ray of light caused by its passage through a gradient of refractive light.

β the angle between the optical axis and the direction at which a ray of light enters a solution.

ε extinction coefficient.

θ the angle between the schlieren analyser and the x direction; also used for the angle subtended by the cell of an ultracentrifuge at the centre of rotation.

ϕ the angular deviation of a ray of light in a calibration cell.

λ the wavelength of light in vacuo.

Subscripts

b   boundary

r   reference point

1

INTRODUCTION

THERE are many instances in physical biochemistry where we wish to measure the distribution of the solute in a solution without disturbing the solution in any way. These include, especially, measurements of the sedimentation of macromolecules in the ultracentrifuge, their migration under free electrophoresis, and free diffusion. In the case of the ultracentifuge it is obvious that we cannot make any direct physical contact with the solution under study (it is in a rotor spinning at anything up to 1100 revolutions per second), but even in free-boundary electrophoresis and diffusion experiments it is impossible to make physical contact with the solution without disturbing it. So direct analytical procedures are not possible. (In some cases, sedimentation work is carried out in preparative machines and the distribution of the solute is determined after the centrifuge is stopped by collecting samples from a hole punched in the bottom of the tube, which are analysed by any convenient means; this book is not concerned with this technique.) The only way of making continuous measurements while the experiment is in progress is to shine a beam of light through the solution and use some optical property of the solute to determine its distribution from the emergent beam.

There are three optical properties of a solvent which may be modified by the presence of a solute. These are absorption, refraction, and rotation of the plane of polarized light. Of these the last is not the basis of any common optical system for measuring distributions of concentration and will not be discussed further here.

Absorption

The absorption of monochromatic light by a solution of an absorbing solute in a transparent medium is described by the familiar Beer–Lambert law which may be written

$$\log (I_0/I) = \varepsilon ca, \tag{1.1}$$

where I_o is the intensity of light incident on the solution, I is the intensity of light transmitted by the solution, c is the concentration in any convenient units, a is the path length through the solution, and ε is a constant called the extinction coefficient whose value depends on the nature of the absorbing solute and on the units in which c is measured. The quantity on the left-hand side of eqn (1.1) is dimensionless and is called the *extinction E* of the solution and, since the light intensities occur only as a ratio, the units in which these intensities are measured are not important. The extinction coefficient ε is normally a true constant over the range of concentrations used in physical biochemistry, but in any case the optical system can only measure the extinction of the solution, and this can always be related to the concentration in any desired units if the value of ε and, if necessary, the variation of ε with c are known from other measurements.

The Beer–Lambert law applies strictly only to monochromatic light. If the light illuminating the cell consists of a broad band of wavelengths, and especially if this band is centred on the edge of the absorption band of the solute, the law will fail, the deviations being greatest at high values of the extinction. However, if the light is composed of a band of wavelengths which is very narrow compared with the width of the absorption band of the solute, the law will be obeyed within the range normally measured ($0 < E < 2$).

Refractive index

When light passes through a transparent medium, the electrons within that medium oscillate in sympathy with the oscillating electric field of the light wave. If the wavelength of the incident wave is not within an absorption band of the medium, no net absorption of energy from the light occurs. However, these oscillating electrons emit radiation of the same frequency as the incident wave and when this combines with the incident wave it produces a combined wave which, in general, moves more slowly than would be the case in the absence of the oscillating electrons. The ratio of the velocity of the wave in a vacuum to that in the medium is called the *refractive index* of the medium and is almost always greater than unity. (Although it is possible for a medium to have a refractive index less than unity, this never occurs for solutions of macromolecules with which this book is primarily concerned.) Since the frequency of the wave remains constant, this decrease in velocity also results in a decrease in the wavelength so that it is important, when quoting a wavelength, to specify the medium in which it was measured. If not otherwise stated the medium may be taken as a vacuum, where the wave has its maximum velocity and wavelength.

When most non-absorbing solutes are dissolved in a transparent solvent, they cause an increase in the refractive index of the solvent according to the relation

$$n = n_0 + \frac{\mathrm{d}n}{\mathrm{d}c} c, \tag{1.2}$$

in which n and n_0 are the refractive indices of the solution and solvent respectively, c is the concentration of the solute, and $\mathrm{d}n/\mathrm{d}c$ is a constant which, if c is measured in g ml^{-1}, is called the *specific refractive increment*. This increment is a quantity which appears to be independent of the detailed structure within a given class of macromolecules, so that frequently its value can be assumed and eqn (1.2) used to determine the concentration without incurring a large error. For instance, most proteins at a wavelength of 546 nm have specific refractive increments close to 0·183 ml g^{-1}, the nucleic acid DNA has a value of about 0·19, and carbohydrates have values of about 0·14. A few solutes produce solutions which do not follow eqn (1.2) especially at high concentrations. For such systems a more complex form of this equation must be used.

If a ray of light enters a medium of different refractive index at an angle to the surface other than 90°, the change in velocity of the light results in a change in the direction of the ray. Apart from a small effect discussed on p. 69, this is not an important effect in the context of this book because, in all the optical systems described here, the light enters the solutions at 90° to the entrance windows and no refraction takes place at the windows. However, there is another cause of refraction which is most important. Rays of light are deviated if they pass through regions where there is a gradient of refractive index even though they enter at 90° to that gradient. This effect was first described by Wollaston in 1800, and the reason for it is most readily understood in terms of wavefronts.

A wavefront is a surface which joins corresponding positions on the rays which constitute a beam of light. It moves forward at the speed of light in a direction at right-angles to itself. Thus a point source of light emits expanding

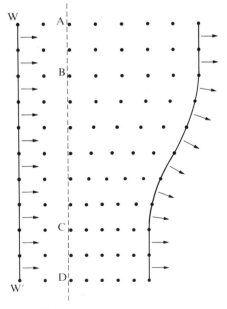

FIG. 1.1. The bending of a beam of light caused by its passage through a gradient of refractive index. For a full description see the text.

spherical wavefronts and parallel light has plane wavefronts. Fig. 1.1 shows a cell, with infinitely thin windows, containing a solution whose refractive index is constant over the region A–B, increases over the range B–C, and remains constant at the higher level over the region C–D. WW' represents a plane wavefront approaching the cell. When this wavefront reaches the cell its velocity is reduced by the higher refractive index, this decrease being least in the region A–B, intermediate in the region B–C, and maximum in the region C–D. Consequently, when the wavefront emerges, it is distorted as shown. The wavefront advances at right-angles to itself as shown by the arrows, which also indicate the directions of the emerging beams of light. The rays have therefore been refracted in the region where there is a gradient of refractive index by an amount proportional to and in the direction of that gradient. This effect is used in the schlieren optical system described in detail in Chapter 3.

The decrease in the velocity of the light as it passes through a region of higher refractive index results in an apparent increase in the optical path as measured in wavelengths. It is this effect which is used in the interference optical system described in detail in Chapter 4. It will be convenient to refer to the *optical path length*, meaning the distance between two points measured in terms of the wavelength of the light in the medium concerned. It is equal to the physical path length multiplied by the refractive index of the medium.

The three dimensions in space

In the discussion of the various optical systems it will be necessary to distinguish the three dimensions in space. The direction in the cell in which the changes in concentration occur and in which any externally applied field acts is called x (Fig. 1.2(a)). The optical axis is called y, and the direction at right-angles to both of these is called z. Such terms as 'top', 'bottom', 'up', 'down' always refer to the x direction with the external force acting 'downwards'. It is the normal convention, however, to draw the final diagrams with the x direction horizontal and the z axis vertical (Fig. 1.2(b)). This results in the slightly confusing situation that 'vertical' in the cell means 'horizontal' on the pattern. However, this convention is now so universal that it will be retained here. It is usual for the patterns to be shown with the top of the cell to the left (and is so on all the diagrams in this book), but occasionally in the literature the patterns are printed the other way round (i.e. 'up' to the right). In sedimentation patterns the meniscus will always indicate the top of the cell, and in electrophoresis the direction of migration is usually indicated. It is a result of this convention that 'downwards' in an ultracentrifuge cell means 'outwards from the centre of rotation', and, since modern ultacentrifuges spin the rotor about a vertical axis, this is not the same direction as the earth's gravitational field.

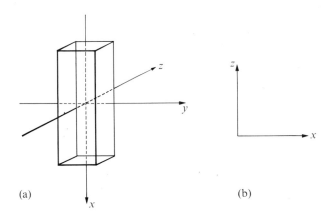

FIG. 1.2. The convention adopted in this book for the labelling of the three dimensions in space, (a) in the cell and (b) in the diagrams. y is the optical axis and x is the direction in which migration occurs.

In Chapter 6, which is concerned with the ways in which measurements of the photographic records are converted into the required distances and concentrations within the cell, the measurements on the photographs will be designated by the corresponding capital letters X, Z, whilst the equivalent distances in the cell will be indicated by the lower-case letters x, z. Because of the convention of showing diagrams with the x direction horizontal, X is a horizontal measurement on the photograph even though it is related directly to a vertical measurement in the cell.

2

ABSORPTION

Spectra

MOST macromolecules of biological interest have optical absorption bands somewhere within the observable spectrum. This wavelength range on most spectrophotometers spans from about 180 nm in the far ultraviolet to about 1000 nm in the near infrared. Special infrared spectrophotometers extend this range up to about 15 μm, but such measurements are not commonly used for studies of solutions because most solvents also absorb infrared light. No optical system currently in common use covers all of this wide range, and the range of about 230 nm to 600 nm may be taken as the generally useful one. Within this range many of the macromolecules of interest to biologists and chemists have absorption bands. For instance, almost all proteins have a band centred at about 280 nm with a very much more intense band at about 195 nm. So intense is this latter band (it is caused by the peptide bond itself) that even on the edge of the band at 230 nm the extinction can be three or four times greater than that at 280 nm. The nucleic acids have a similar absorption band centred about 260 nm with a much more intense absorption below 230 nm. On the other hand, some proteins of particular interest have absorptions in the visible, a familiar example being haemoglobin, which has strong absorptions in the green (538 nm, 566 nm) and blue (416 nm) parts of the spectrum and also in the nearer part of the ultraviolet (345 nm) as well as the typical protein bands mentioned above. Thus, with the exception of the carbohydrates and some of the simpler synthetic polymers, most macromolecules show an absorption somewhere in the observable range. For those that do not, it is often not difficult to introduce absorbing groups into the molecules by chemical modification.

It should be remembered, in the context of experiments with absorption optical systems, that the absorption bands of molecules in solution are very broad, and wavelengths close to those giving maximum absorption need not be used. For instance, most proteins show appreciable absorptions at all accessible wavelengths below 300 nm. However, in this case it is important that the illuminating light be of narrow bandwidth, as usually obtained from monochromators. Many simple filters (even the cheaper interference filters) pass light with a considerable bandwidth. If this band is centred on the edge of the absorption band of the absorbing substance, the Beer–Lambert law (eqn (1.1)) may not be obeyed. The extent of departure from this law depends upon the bandwidth of the light in comparison with the width of the absorption band of the chromphore, but may be quite considerable at high values of the extinction.

The optical system

This is the simplest of the optical systems, at least in principle. The cell containing the absorbing solution is illuminated with light of a suitable wavelength and photographed with a simple camera. More recently complex photoelectric detectors have replaced the photographic film as the final detector with considerable advantages.

The first requirement is a source of monochromatic light of a suitable wavelength. For various technical reasons, which need not concern us here, this light source usually has to be more intense than those used in spectro-photometers. The earliest and simplest arrangement was a simple gas-discharge lamp and a set of filters to isolate the relevant line in the emission spectrum of the lamp. A very common combination was a medium-pressure mercury-arc lamp combined with filters of chlorine and bromine. These filters isolated the mercury lines at 248 nm, 254 nm, and 265 nm, together with a broad band of light at wavelengths greater than 570 nm. By using a photographic film which was only sensitive to light of wavelength less than 540 nm, one had a system ideal for the study of the nucleic acids. With modern interference filters this kind of system could be a simple and con-venient one if measurements at only a very few wavelengths are required.

If measurements at a large number of different wavelengths are likely to be needed, then a monochromator is required. These are essentially the same as those used in spectrophotometers, but with the addition of more intense light sources, and may be built into the optical system or simply added in place of the simple light source. High-pressure, xenon-arc lamps are the most suitable sources because they are powerful and approximate well to point light sources. The addition of mercury to the lamp increases the emission in the region of the mercury emission lines. This can be useful but has the disadvantage of causing somewhat uneven illumination of the exit slit of the monochromator when set to wavelengths on the edge of one of these emission lines, which are considerably broadened by the high pressure within the lamp.

The optical system is essentially very simple (see Fig. 2.1). The light source (or the exit slit of the monochromator), which should ideally be a point source, is placed at the first focal point of a collimating lens L_1, which produces a parallel beam of light to illuminate the cell. This is necessary to ensure that the cell is evenly illuminated over its whole length. The only condition on L_1 is that its diameter is greater than the longest edge of the cell. Since in practice point light sources cannot be made, the beam illuminating the cell is not strictly parallel, but if the source is evenly placed about the focal point of the lens the cell remains evenly illuminated for all practical

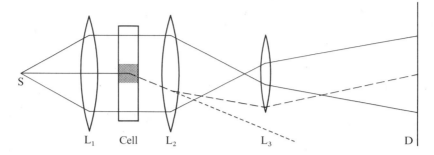

FIG. 2.1. The layout of an absorption optical system. The shaded part of the cell contains a gradient of refractive index, and the light that is deflected by this gradient is shown by the dashed line. The dotted line indicates the path this ray would have taken in the absence of the condensing lens. S = source; L_1 = collimating lens; L_2 = condensing lens; L_3 = camera lens; D = detector.

purposes. Since the focal length of the lens varies with wavelength, it is strictly necessary to adjust its position relative to the source for each wavelength required. In practice, however, a mean position is usually chosen and the unevenness of the illumination tolerated.

The size of the cell is partly dictated by considerations other than the optical system. The length will be chosen according to the resolution required by the experiment. The width in static experiments is not important but in the ultracentrifuge it determines the fraction of the incident light which passes through the cell even when the solution is fully transparent. This arises because the cell, as it rotates in the rotor, only passes through the optical system once in every revolution so that, averaged over a time which is long compared with the rotation period of the rotor (which may be only 1 ms), the light intensity is reduced by a factor of $360/\theta$, where θ is the angle (in degrees) subtended by the cell at the centre of rotation of the rotor. The most important dimension as far as the optics are concerned is the depth a in the direction of the beam. The larger this depth the greater is the sensitivity since, for a given concentration of absorbing solution, the extinction increases with this depth (eqn (1.1), p. 1). However light from the cell is later to be focused by a camera and the larger the depth of the cell the more difficult it becomes to focus it accurately. In practice depths no greater than 30 mm are used.

The only other essential optical component is the camera lens L_3 which focuses an image of the cell on to a detector D. In practice, another lens L_2, known as the condensing lens, is often placed immediately after the cell. The purpose of this lens is to ensure that all the light which passes through the cell reaches the camera lens. At best the emerging beam is parallel so that a camera lens at least as big as the cell is required and, if the whole optical system is not to be too long, it has to be of fairly short focal length which would make it an expensive item. More important than this, however, is that the rays may not pass straight through the cell. If gradients of refractive index exist in the solution, the rays of light will be refracted and may miss the camera lens altogether if it is not very large. The effect of the condensing lens is to reduce the effect of these deviations of the light and so to ensure that most of them pass through the camera lens and reach the detector at D. This effect is illustrated by the dotted line in Fig. 2.1. The condensing lens also has the effect of reducing the spherical aberration of the camera lens by ensuring that all rays pass through it close to its axis. One manufacturer (M.S.E.) omits this lens and uses a large and fully-corrected camera lens with no untoward effects.

Another manufacturer (Beckman) has a design in which the camera lens is omitted and replaced by a concave mirror. The advantage of this is that the focusing of the system is the same at all wavelengths (apart from a small effect of the condensing lens), unlike the lens system which has to be refocused for each different wavelength. However M.S.E., who do not use a condensing lens, provide interchangeable achromatic camera lenses, one of which covers the ultraviolet and the other the visible range, and the only change required is to take one lens out and put the other in (they are mounted on a rotating carriage) at one particular wavelength. They claim that no other focusing is necessary when changing wavelength.

The detector
Photographic plates

The simplest form of detector, which has been used for many years, is the photographic plate. The camera lens focuses the image of the cell on to the plate, which records the intensity of the light passing through the cell at each level. An illustration of a picture taken this way is given in Fig. 2.2(a). In conformity with normal practice this picture is turned on its side so that the top of the cell is on the left and the bottom on the right. Thus downwards in the cell is left-to-right in the picture. The picture is a negative, so that it is blackest where the most light has passed through the cell and lightest where the maximum absorption has occurred. The picture was taken when DNA was sedimenting in an ultracentrifuge, and the image of the cell is the long image in the middle of the picture. The dark patches at the ends are formed by light passing through holes in the rotor and provide reference marks. The meniscus is visible as a fine white line near the left-hand side of the cell image to the right of which is a dark area where the solution was transparent. Further to the right (that is further down in the cell) the image becomes lighter and this is the boundary of the DNA solution which absorbs the light.

In order to convert the photographic image into a convenient quantitative form, the film is scanned by a microdensitometer. This automatically plots

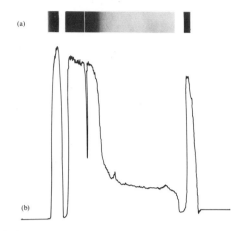

FIG. 2.2. (a) A typical photograph obtained with the absorption optical system of an ultra-centrifuge, and (b) the trace obtained from this photograph by the use of a microdensitometer.

the extinction of the film as a function of the distance across it. The trace of Fig. 2.2(a) is shown in Fig. 2.2(b). It is a property of photographic emulsions that the extinction of the developed plate is a linear function of the logarithm of the light intensity so long as the latter exceeds a certain value. Thus, if this value is exceeded at all points, the curve drawn by the densitometer is an inverted plot of the extinction of the solution as a function of the position in the cell. This point will be discussed more fully in Chapter 6.

Photomultipliers

More recently many workers using absorption optics on the ultracentrifuge have replaced the photographic film by a photoelectric detector, and these together with their associated electronic circuits have become known as *scanners*. In their simplest form they do little more than replace the photographic film, but the first description of a more highly developed form was given by Lamers, Putney, Steinberg, and Schachman (1963) and another form which differed in several points of detail, but which worked on the same general principles, was described by Spragg, Travers, and Saxton (1965). In each case the rotor of the centrifuge carries two cells, one containing the solution under study and the other a suitable reference solvent. (In the design described in the earlier of these papers the two solutions were contained in two parts of a double cell similar to those used with the Rayleigh interferometer (see Chapter 4); this is no different in principle from two distinct cells but alters the details of the electronic circuits.) As the rotor turns, first one cell then the other passes through the optical system and its image is focused on to a screen containing a fine slit. The light that passes through the slit falls on to a photomultiplier which gives out an electric current proportional to the intensity of that light. Since the photomultiplier gives out a current only when the cell is actually passing through the optical system, this current consists of a series of very short pulses, the lengths of which depend upon the speed of the rotor and on the width of the cell in the rotor and can be as short as 6 μs.

The output pulses from the photomultiplier are passed to electronic circuits which are shown in block form in Fig. 2.3. After amplification in a linear pre-amplifier, the pulses are passed through an amplifier which has a

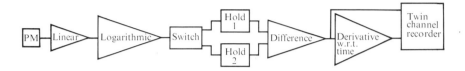

FIG. 2.3. A block diagram of the electronic circuits of a scanner using a photomultiplier (PM) as the light detector. ▷ represents an operational amplifier whose output is related to its input by a function indicated by the word inside the symbol.

logarithmic response, so that the pulses then have amplitudes proportional to the logarithm of the amplitudes of the pulses from the photomultiplier and hence to the logarithm of the intensity of the light striking the photo-multiplier. These pulses are then passed through an electronic switch which routes them to one of two holding circuits. The pulse derived from light passing through the reference cell is routed to holding circuit 1 (Fig. 2.3), where its peak amplitude is held to await the arrival of the second pulse when the cell containing the experimental solution is brought by the rotor into the optical system. This second pulse is fed by the switch to holding circuit 2. The peak amplitudes of these two pulses are then fed to a difference amplifier where they are subtracted, and the difference is a current proportional to the extinction of the test solution at the particular level in the cell from which the light has come before passing through the slit over the photomultiplier. The output from the difference amplifier is fed to a recorder and also to an amplifier which gives an output proportional to the first derivative of its input with respect to time, and this output is fed to a second pen on the recorder. Because the scanning occurs at a constant speed, this derivative with respect to time represents the derivative of absorption with respect to distance in the x direction of the cell.

The slit with the photomultiplier is then moved across the image of the cells at a slow, constant speed (or in the design of Spragg et al. (1965) the images are moved across the photomultiplier slit by a moving mirror) so that the system measures the extinction of the solution at successive levels down the cell, and the recorder records this and its derivative. Thus the two traces on the recorder are of the extinction E and the first differential dE/dx

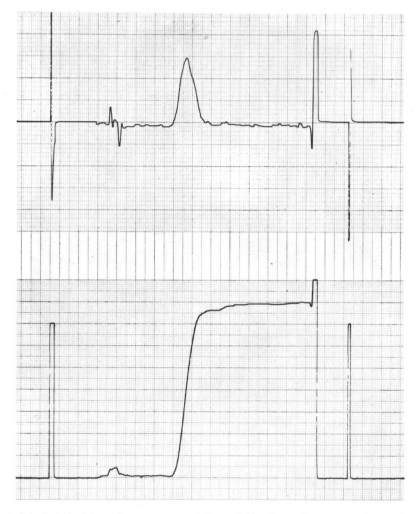

Fig. 2.4. A typical trace from a commercially-available electronic scanner using a photo-multiplier. The lower trace is of E versus x, and the upper trace is of dE/dx versus x.

as a function of x. A typical example of a trace obtained from this type of scanner is shown in Fig. 2.4.

Both these systems have been developed commercially and are available from two manufacturers of analytical ultracentrifuges for fitting to their machines†. Each has been further developed to allow up to five samples together with five reference solutions to be run in the rotor simultaneously.

† A design based on that by Lamers *et al.* (1963) is available from Beckman Instruments, and one based on that by Spragg *et al.* (1965) is available from M.S.E.

The additions to Fig. 2.3 to allow for this are fairly simple. The electronic switch becomes more complicated and has ten output channels, any pair of which can be fed by means of mechanical switches to the holding circuits and hence to the difference amplifier. Consequently the absorption of any one cell can be compared with that of any other so that up to nine samples could be run simultaneously with a single reference solvent. This part of the system is known as a *multiplexer*, and, as well as the additional electronic circuits mentioned above, they require some kind of sensor to indicate the position of the rotor to ensure that the correct cell is being scanned at all times.

An interesting variation of this type of scanner for the ultracentrifuge has been designed by M.S.E. for one of their analytical centrifuges (the Centriscan), and is shown in Fig. 2.5. Light from a mercury-arc lamp is focused by a

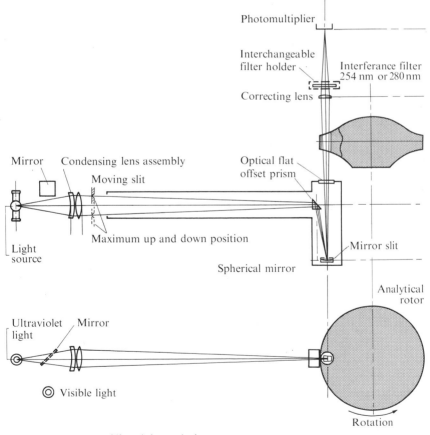

Ultraviolet optical system

FIG. 2.5. The layout of the absorption optical system of the Centriscan.

condensing lens on to a spherical mirror from whence it passes through the cell to a photomultiplier. Close to the condensing lens is a diaphragm with a narrow slit in it which is focused by the spherical mirror on to the cell. This diaphragm is moved slowly downwards so that the narrow beam of light slowly traverses the cell. The photomultiplier receives all the light that passes through the cell, the scanning being achieved by the moving beam illuminating the cell. A simple modification enables the system to record derivative refractive-index curves and this will be described in Chapter 3 (p. 36).

Television camera

A scanner based on rather different principles has been described more recently (Lloyd and Esnouf 1974) in which the detector is a vidicon television camera. The scanning of the images is done with a beam of electrons which means not only that there are no moving parts but also that the scanning process is extremely rapid. The whole cell image is scanned 50 times per second compared with once in 6 s or more for the mechanical scanning systems.

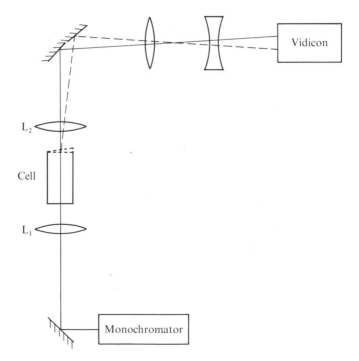

FIG. 2.6. The optical system of a scanner which uses a television camera for detecting and scanning the images.

The vidicon tube is less sensitive to light than a photomultiplier, but this is largely, if not entirely, offset by two of its features. First, the image of the cell can be about half the size of that presented to the photomultipliers without loss of resolution, and this results in a fourfold increase in the intensity of the image. Secondly, the tube does not respond to the instantaneous light intensity striking it at any moment but to the total amount of light which strikes any given point on its photoconductive target between successive passages of the scanning beam of electrons across that point. In this respect the tube resembles a photographic film; the longer the time of exposure the greater is the output signal. At normal television speeds the target is scanned 50 times each second, but this interval of 20 ms can be increased to 2 s or more, resulting in at least a hundredfold increase in sensitivity. Thus the tube does not have to respond in the very brief moment when a cell passes through the optical system in the ultracentrifuge, but it gathers the light from all parts of the cell each time the rotor comes round and stores it for readout from time to time by the scanning beam of electrons. As a result of this property of the vidicon tube, the system cannot use the rotor itself as the beam splitter for double-beam operation, and a rather more complex optical system has to be used (Fig. 2.6), but this also means that the system in operation is completely independent of the rotating rotor and that it could be used without modification on any apparatus where extinction had to be measured as a function of one linear coordinate, such as free-boundary electrophoresis, or diffusion experiments, or in the flow-type, fast reaction-kinetics apparatus. All that is necessary is to project the images of the solution and reference cells side by side on to the face of the vidicon tube, and in static apparatus this can be done with a much simpler optical system than has to be used in the ultracentrifuge. In the ultracentrifuge the two cells are distinguished by fitting one with a prismatic window which deflects the light in a direction at right-angles to the radial direction. The normal condensing lens produces, therefore, two images of the light source at its focal point, one from light passing through the cell with plain windows and the other from light passing through the cell with the prismatic window. These images are arranged to fall on two mirrors inclined at a very small angle to each other so that, after passage through the camera lens, the light forms two images side by side on the face of the vidicon tube. The inclusion of the mirrors, which bend the optical system, is not as much of a disadvantage as might at first appear, since all ultracentrifuges have optical systems bent in this way because their rotors rotate about a vertical axis. The two mirrors simply replace a single mirror which was there in the simple photographic system. In the design described by Lloyd and Esnouf (1974) the camera lens consisted of a telephoto combination of positive and negative lenses because the image was required to be slightly smaller than the cell and a single lens would have had to be placed in an inconvenient position. The use of two lenses here had the further advantage over a single lens that the magnification of the image could be kept constant at all wavelengths by suitable movement of both lenses.

The electronic systems are shown in block form in Fig. 2.7. The only parts which have to be mounted in the centrifuge itself are the vidicon tube with its associated scanning coils and the first stage of linear amplification. The rest of the circuits can be remote from the machine and connected to it by a single 25-way cable which can be as much as 50 ft long. The images of the cells are

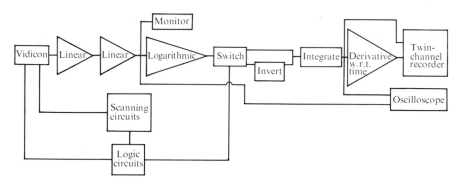

FIG. 2.7. A block diagram of the electronic circuits of a scanner using a television camera. ▷ represents an operational amplifier whose function is indicated by the word within the symbol.

arranged to be at right angles to the line-scan of the camera, so that the output from the vidicon camera is essentially the same as that from the photomultipliers described above; that is a pair of pulses on each line of the television scan, one corresponding to the solvent cell, the other to the solution. The analogue measuring circuits, which derive the extinction of the solution from these pulses, are essentially very similar to those described previously. The main difference is that the pulses are not stored in a holding circuit which stores their peak amplitude but in an integrator which stores the product of the amplitude and the duration of the pulses. This is possible because the widths of the pulses are controlled not by the speed of rotation of the rotor (as with the earlier systems) but by the television scanning and the widths of the images, which are in turn determined by a mask over the face of the vidicon tube. Integration has the advantage over peak detection that it is less affected by any electronic noise in the system. The electronic switches are also different in that they are normally open (preventing the signal from reaching the integrator) except when closed by the logic circuits, and this happens only for the exact moment when a pulse arrives from the camera. This is possible because, again, the timing of the pulses is fixed by the television scanning and not by the rotating rotor. This has the advantage that the system is immune to noise during the intervals between pulses, and it is at this very time that the logarithmic amplifiers have their highest gain and any noise in the system is greatly amplified.

The scanning circuits are very similar to those used in commercial television and their purpose is to cause the scanning beam of electrons in the vidicon to sweep across from one image to the other (the lines of the television scan) whilst also moving it slowly down the images so that in a period of 20 ms the whole image is scanned. The logic circuits have three important functions which are not present in the photomultiplier systems. First, they control the operation of the electronic switches, closing them only when a pulse is expected from the camera and routing the first pulse directly to the integrator and the second to the inverter so that it is subtracted from the first in the integrator. Secondly, they control the electronic intensification of the image. As was mentioned above, the output from the vidicon tube for a given light intensity depends upon the interval between successive passages of the electron beam across a given point on the photosensitive target. The logic circuits do this by switching off the electron beam in the tube so that it fails to strike the target for periods which may be selected manually from 20 ms to 4 s. Thirdly, they operate a slow-scanning system. The normal scanning system of the camera scans the whole picture in only 20 ms, and,

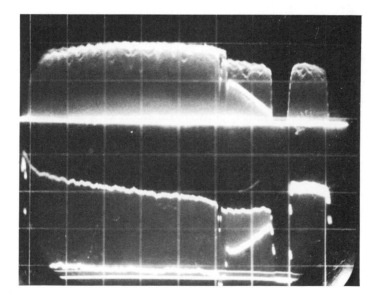

FIG. 2.8. A photograph of the output of the television scanner as seen on the oscilloscope screen. The top trace is the signal from the camera itself, and the bottom trace is the output of the analogue measuring circuits. The picture shows a protein at sedimentation-equilibrium, and the solution fills only the bottom third of the cell. The apparent fall in the signal from the camera at the top of the cell is not real but is caused by the edge of the oscilloscope screen.

although the analogue measuring circuits can respond in this time and the E versus x and dE/dx versus x curves can be displayed continuously on an oscilloscope (Fig. 2.8), no recorder can make a permanent record in such a short time. It is necessary to slow down the readout so that a recorder can respond. This is done by the simple process of reading out just one line during each frame of the television scan and stepping this line down one by one each frame. Since there are 625 lines to the frame and there are 50 frames per second, it takes just $12\frac{1}{2}$ s to read out the complete scan. If resolution is not to be lost, the recorder must be able to give a full-scale deflection in 20 ms and this is well within the capability of the type of recorder which uses a narrow beam of ultraviolet light to write on photosensitive paper.

Comparison of detectors

Although photographic films are very cheap, they have several important disadvantages compared with the electronic detectors. They have to be developed and fixed before they can be examined, whereas the photoelectric systems give more instant results. With the television system the images of the cell can be displayed continuously on a monitor screen, which can be very useful for focusing the system and for watching for the sedimentation of aggregates during the acceleration of the rotor to full speed. Because photographic emulsions are made with gelatin, which is a protein, they tend to be less sensitive in the regions of the protein absorption bands, which are the very parts of the spectrum of most interest to the biochemist. For this same reason they become completely insensitive to wavelengths below about 235 nm, although special films are available that have the silver halide grains concentrated in the surface layers of the emulsion and which are more sensitive at these very short wavelengths. Although the sensitivity of most photoelectric detectors falls rapidly at shorter wavelengths they are still useful down to 230 nm or below. It is very difficult, especially in the ultra-centrifuge, to keep all the optical components completely clean and dust and oil particles show up as shadows on the photographic plates. The scanners, however, can easily be made double-beam in operation and the effects of these particles are automatically subtracted out of the traces. In addition variations of the output intensity of the lamp, or unevenness of the illumination across the cell are allowed for automatically.

Results

The majority of absorption optical systems yield their results as plots of the extinction as a function of the distance down the cell. Since extinction is normally a linear function of the concentration, these plots may be taken as plots of the concentration of the absorbing solute as a function of the position in the cell. In addition, the electronic scanners can be made to give an output which is proportional to the first differential of this curve. This has certain advantages. The traces so obtained look just like those obtained by the schlieren system (see Chapter 3) and it is probably easier to locate a boundary by means of the top of a peak than by the mid-point of a sigmoid curve. More important than this is the fact that a differential output shows up over-lapping boundaries very much more clearly than does an integral output. This is illustrated in Fig. 2.9.

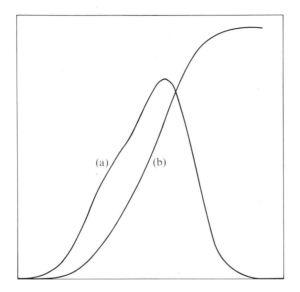

FIG. 2.9. A comparison of the representation of two overlapping boundaries by (a) a differential (dE/dx versus x) trace, and (b) an integral (E versus x) trace. Curve (a) shows the boundaries more clearly than curve (b).

The most important advantage of the absorption system over the others is its ability to discriminate between different solutes. As pointed out in Chapter 1, the refractive increments of most polymeric substances are nearly the same so that the refractometric systems are intrinsically unable to distinguish one solute from another. Although it is true that the absorption bands of many of the macromolecules of interest to a particular worker will be similar, it is always possible to distinguish them by observing at two wavelengths where the ratio of the extinction coefficients differ. An interesting example of this use of the system is the work of Gerhart and Schachman (1965) on the binding of the regulator molecule cytosine triphosphate (CTP) to the enzyme aspartate transcarbamylase and its subunits. By using the bromo derivative of CTP, they were able to use a wavelength (298 nm) where the extinction of the BrCTP was high and that of the protein low and so observe directly on which subunit the BrCTP was bound. The increase in refractivity of the subunit by the binding of BrCTP would be very small but the increase in extinction was very large and so the result was clear and not masked by the contributions from the two subunits.

In many cases the sensitivity of the absorption system is very much greater than the refractometric systems, but this depends upon the molecule under study and its absorption spectrum. For instance the sensitivity of a system working at 260 nm is at least five times more sensitive for the nucleic acids than a refractometric method, but almost useless for a carbohydrate which has no absorption in the observable range. An absorption system operating at a wavelength of 230 nm is nearly one hundred times more sensitive to a protein than is a refractometric method or, indeed, an absorption system operating at 280 nm.

The principle disadvantages of the absorption systems are the delay involved with the photographic detectors whilst the film is developed, and the expense of the electronic scanners.

3

SCHLIEREN

THE schlieren optical systems which are now in common use have been developed from an original system proposed by Toepler in 1866 for detecting inhomogeneities in the glass used for the manufacture of lenses. These inhomogeneities usually appear as long, thin, streamer-like regions of sharply different refractive index from the bulk of the material, and result from incomplete mixing of the glass. The German name for these is *schlieren* from which the optical system takes its name. Since this system detects these schlieren by the deviations they cause in a beam of light passing through them, the name 'schlieren' is applied to any optical system which measures changes in refractive index within a transparent medium by means of the deviations of a beam of light. It is very similar to the Foucault knife-edge test used to detect aberrations in lenses and mirrors (Foucault 1859).

In the discussion which follows it will frequently be necessary to refer individually to each of the three dimensions in space. The reader is referred to the final section of Chapter 1 and to Fig. 1.2, where the three axes are defined.

It was shown in Chapter 1 that, when a beam of light passes through a cell containing a gradient of refractive index in a direction at right-angles

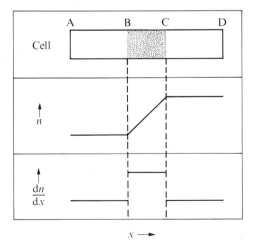

FIG. 3.1. In the optical diagrams in this chapter, the cell is assumed to contain the very special distribution of solute shown here.

to the direction of the beam, it is deviated through an angle proportional to and in the direction of that gradient. The schlieren optical system measures this deviation as a function of the x coordinate, and consequently measures dn/dx as a function of x, where n is the refractive index of the solution.

In order to simplify the diagrams in the discussion which follows, the refractive index within the cell is assumed to have the rather special (and exceedingly unusual) distribution shown in Fig. 3.1. At the top of the cell, from A to B, the refractive index is constant. At B it suddenly starts to rise at a constant rate until point C is reached, where the index suddenly becomes constant again at the higher value. The curves of n versus x and of dn/dx versus x are shown in the figure. In the diagrams the area of the cell between B and C is shaded.

The optical systems
Earlier forms

The schlieren optical system has developed slowly from quite simple forms to its present complexity. The old forms make a fascinating subject for study by the historian, but the account given here is not intended to be a comprehensive historical review of these optical systems. Rather it is intended to describe only those systems that either were in common use at one time or contribute to an understanding of the modern system.

The earliest system was described by Thovert (1902). He derived an equation describing the path of a beam of light through a gradient of refractive index and showed that the angular deviation is proportional to that gradient. He used a horizontal slit as a light source and focused an image of this through the cell. He then measured the deviations of the light directly by observing the movement of this image when light was allowed to pass through only a very small part of the cell. He masked the cell with a screen containing two narrow, closely spaced, horizontal slits. The result was a set of interference fringes in the image of the light source, from which he was able to determine the position of the deflected image. In 1914 Thovert published an improved version based on an idea by Wiener (1893). The cell was illuminated with a parallel beam of light derived from a small light source placed at the focal point of a collimating lens. The cell, which was as wide as it was high, was masked by a screen, in the xz plane, containing a narrow slit at 45° to the vertical. After the cell he placed a cylindrical lens with its axis horizontal. This produced in its focal plane an image which, in the absence of any deviations of the light in the cell, was a horizontal line. In the presence of such deviations the line became bent and the displacement of the line below its original position was proportional to the deflection of the light in the cell. The resulting curve was, therefore, a plot of the gradient of the refractive index as a function of the position in the cell in rectangular coordinates. The representative pictures shown by Thovert (1914) are very good and would do credit to later systems. This system was very simple but illustrates two features which are important in later versions. The first point to notice about the cell and the solution in which the diffusion, or other process, is occurring is that only vertical coordinates are of any significance. Gradients of refractive index are usually accompanied by gradients of density, and consequently any horizontal gradients would be gravitationally unstable. The second point is that the deviation of the light occurs in the vertical direction, which is the same direction as is of interest in the cell. Consequently, if a plot such as that produced by Thovert (1914) or by the modern system is to be obtained in rectangular coordinates, either the deflection or the cell coordinates must be rotated through 90°. In Thovert's system the cell coordinates were rotated by means of the slit at 45° which gives a 1:1 correlation between the vertical and horizontal coordinates in the cell.

Although the Thovert system was used by other workers (e.g. Tanner 1927) it was superseded by a more accurate, although more laborious, system described by Lamm (1937). Lamm mounted a transparent scale close to but not coincident with the cell, and photographed it through the cell. The image of the scale was distorted by gradients of the refractive index in the cell, and by careful measurement of this distortion it was possible to deduce the variations of dn/dx with x within the cell. Rather long calculations were needed to convert the measurements into the required results and, since it was not possible to focus both the scale and the cell simultaneously on to the photographic plate, the system could not resolve the x coordinate accurately.

A much simpler way of locating the maxima in the gradient curves was described by Tiselius, Pedersen, and Eriksson-Quensel (1937). The source was a horizontal slit, light from which was rendered parallel by the collimating lens L_1 (Fig. 3.2). The horizontal slit acts as a point in the vertical plane so that in this direction (the only one that matters in the cell) the light passing through the cell is parallel. After passing through the cell, the light

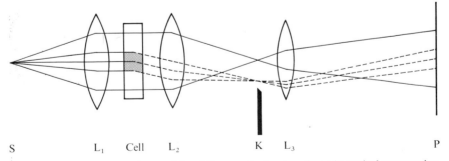

S L_1 Cell L_2 K L_3 P

FIG. 3.2. One of the earlier forms of the schlieren optical system shown in vertical cross-section. The source S is a horizontal slit, and K is a knife-edge parallel to S and which can be moved from the position shown towards the optical axis.

passes through the condensing lens L_2, which forms an image of the light source in its focal plane. Light passing through the parts of the cell with no gradient of refractive index (A–B and C–D) forms an image on the optical axis. However rays of light which are deviated by passing through such a gradient in the cell produce an image displaced from the optical axis (these rays are shown in Fig. 3.2 by dotted lines). It is important to note that light which suffers a given deviation in the cell will form an image of the light source at a given position no matter where in the cell the deviation takes place. The light rays then pass through a camera lens L_3, which forms an image of the cell on a photographic plate or ground-glass screen P.

So long as the deviated rays strike the camera lens they are brought back by the lens to their correct relative positions in the image. Consequently the image appears uniformly bright and contains no information concerning the gradients of refractive index within the cell. However, in the focal plane of the condensing lens, where the images of the light source are produced, there is, in the positions of these images, all the necessary information concerning the magnitudes of any gradient of the refractive index in the cell but no information as to where in the cell these gradients are to be found. However, if an opaque knife-edge K is introduced, as shown in Fig. 3.2, and gradually moved nearer the optical axis, it will eventually begin to intercept the rays making up the most deviated image. When it does so a dark patch appears on the image at P at those parts of the image corresponding to the parts of the cell where the maximum gradients of concentration exist. In the special situation represented in the cell shown in Fig. 3.1, of course, the area B–C will go black for any position of the knife-edge off the optical axis by less than the distance of the deflected image. In a real situation, however, where the gradient of refractive index rises smoothly to a maximum value and then falls smoothly again to zero, a whole series of images is produced in the focal plane of the condensing lens, and a position of the knife-edge could be found where only the most deviated light was cut off and only a narrow black line would appear in the image at P. This will locate the position of maximum gradient with great accuracy. However if there were other maxima in the refractive-index gradient curve (due to there being more than one boundary) any smaller maxima would be missed. If the knife-edge is moved in further, eventually the next greatest maximum in the refractive-index gradient curve shows up, but by then the first black line will have spread into a dark patch, and if there were two overlapping boundaries their resolution may have been lost altogether. This is illustrated in Fig. 3.3. Whatever the position of the knife-edge there is always the possibility of either missing a small boundary or not resolving two large ones. In addition, the gradients of refractive index tend to decrease with time so that constant adjustment of the knife-edge is needed. This system is still available from one manufacturer (M.S.E.), who claims that it is capable of detecting much smaller gradients of refractive

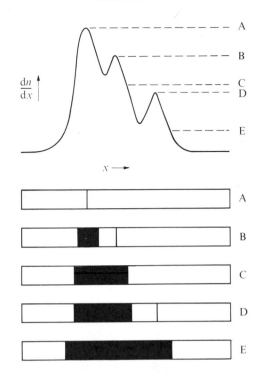

FIG. 3.3. A–E show the types of picture that would be obtained from the system of Fig. 3.2 in the presence of a boundary system shown at the top of the diagram by means of its corresponding dn/dx plot. At A the knife-edge is furthest from the optical axis and only the largest peak is detected. As the knife-edge is moved inwards it eventually detects the second peak at B, but the position of the first peak is now uncertain. At C the resolution of the first two peaks has been lost before the third peak is detected at D. Finally at E the knife-edge is very close to the optical axis and the resolution of all three peaks has been lost.

index than the astigmatic camera to be described below. It is a slight modification of this system that enables an absorption system to record gradients of refractive index (see p. 36).

An improvement of this system was introduced by Longsworth (1939). The knife-edge was moved in towards the optical axis at a steady speed and, at the same time, the photographic plate was moved sideways (into the paper in Fig. 3.2) behind a narrow slit. The result was a series of pictures with the knife-edge at gradually closer positions to the optical axis. The resulting picture was very similar to that produced by the astigmatic camera and knife edge shown in Fig. 3.8(d). A picture obtained with this system is shown by Longsworth (1959). The edge between black and white areas represents a plot of the gradient of the refractive index as a function of the position in the cell. Although this improvement ensured that all the boundaries were properly resolved, the scan was rather slow and, although this was not important in electrophoresis experiments because the electric current could be turned off during the scan, it could be important in sedimentation work in which the boundary might move appreciably during the exposure.

The modern form

The modern form of the schlieren system was devised independently by Philpot (1938) and Svensson (1939). It is optically rather more complex than the earlier forms, but it produces a graph of dn/dx versus x directly without any moving parts. There are two points to be noticed about the system shown in Fig. 3.2. First, as mentioned above, the information concerning the deviations of the light is contained in the positions of the various images in the focal plane of the condensing lens at K, whereas the cell is imaged at P. The problem is to transfer the information at K to P without losing the imaging of the cell at P. The second point is that the image of the cell at P is two-dimensional, whereas only the x direction is of any interest. Thus so long as the imaging of the cell at P is retained in the vertical plane (the x direction) it is of no consequence if focusing of the cell is lost in the z direction. This raises the possibility of using an astigmatic camera containing a cylindrical lens, which will focus the cell at P in the x direction and the plane at K in the z direction. Unfortunately, the deflections of the images at K are in the x direction, and so the first requirement is that this vertical displacement be converted into a horizontal displacement. This was achieved in one of the earliest forms of the optics by introducing at K an opaque diaphragm containing an inclined slit. This is shown in Fig. 3.4 together with the images of

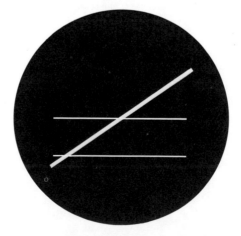

FIG. 3.4. One form of the schlieren diaphragm is a narrow slit placed at an angle to the images of the light source projected onto it. Only two images are shown here but in a real case there would be a very large number.

the light source (which is a horizontal slit) formed by the condensing lens in the presence of the cell depicted in Fig. 3.1. The undeviated image falls across the centre of the slit and light passes through on the optical axis. The light from the deviated image, however, passes through the slit at a point which is not only deviated downwards but also sideways. Thus the presence of the diaphragm with the inclined slit, which is called the schlieren diaphragm, has caused the vertical displacement of the images at K to be converted into a horizontal displacement. (The function of this slit is essentially the same as that of the inclined slit used by Thovert in 1914.) The fact that the vertical displacement remains is of no consequence as will be seen below.

This then is the basic system devised by Philpot (1938) and Svensson (1939). It is essentially a three-dimensional system and so it is difficult to present it in two-dimensional diagrams. However, Fig. 3.5 shows the system in vertical cross-section, Fig. 3.6 shows it in horizontal cross-section, and Fig. 3.7 shows a model of the system constructed by the author. Fig. 3.5 is essentially the same as Fig. 3.2 apart from the inclusion of the schlieren diaphragm and of the lens L_4, which is a cylindrical lens with its axis in the x direction (vertical with respect to the cell), so that in this diagram it appears as a parallel-sided block of glass. It should perhaps be pointed out at this stage that the positions of L_3 and L_4 with respect to one another will differ in different designs. In some designs the cylindrical lens is nearer the schlieren

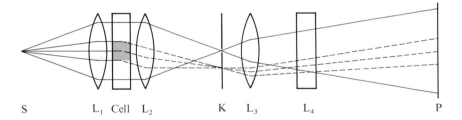

S L_1 Cell L_2 K L_3 L_4 P

FIG. 3.5. The modern form of the schlieren optical system shown in vertical cross-section (the xy plane).

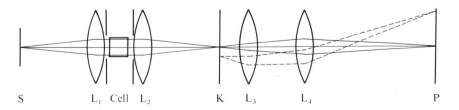

S L_1 Cell L_2 K L_3 L_4 P

FIG. 3.6. The modern form of the schlieren optical system shown in horizontal cross-section (the yz plane).

diaphragm than is the camera lens. The two lenses should really be thought of as a single compound astigmatic lens, but it is convenient for the discussion to keep them separate and, indeed, they are normally two distinct components in the optical system. Also in one design the functions of the condensing lens, L_2, and the camera lens L_3, are combined into a single compound lens.

Referring to Fig. 3.5, the light source, which is a narrow horizontal slit, appears in vertical cross-section as a point. The light is rendered parallel by the collimating lens L_1 and illuminates the cell. After passing through the condensing lens, L_2 it is brought to focus on the schlieren diaphragm at K. The deviated rays produced by the gradient of refractive index are shown dotted, and produce displaced images on the schlieren diaphragm. After passing through the slit in the schlieren diaphragm they pass through the camera lens L_3, which forms an image of the cell at P. On the way, the light passes through the cylindrical lens L_4, which in this plane is simply a block of glass and does not deflect the light, and a true image (in the vertical plane only) is formed on the plate at P.

In horizontal cross-section (Fig. 3.6) the source appears as a line. The cell is normally smaller in this direction than in the vertical direction and so diaphragms are included to prevent light passing round the sides of the cell. The deviated rays cannot be shown in this diagram until they emerge from the schlieren diaphragm when they do so displaced horizontally. After passing through the camera lens the light reaches the cylindrical lens, which

FIG. 3.7. A model of the complete schlieren optical system constructed by the author. The source slit is nearest to the camera and the pattern produced by a typical boundary system is shown on the screen at the far end.

in this plane is a perfectly ordinary lens. The combined action of the two lenses, L_3 and L_4, is to focus the schlieren diaphragm on to the plate at P. Since the deflected rays emerge from the schlieren diaphragm off-axis they are focused at P off-axis. Thus a deflected ray is focused in the x coordinate in the position corresponding to the position in the cell from which it came, and in the z coordinate at a position proportional to the deflection the ray underwent during its passage through the cell. The result is a graph of dn/dx (and hence dc/dx) versus x. For a real boundary this is a bell-shaped curve, which can be seen on the screen of the model shown in Fig. 3.7. A real photograph is reproduced as the frontispiece to this book.

It is evident from Fig. 3.4 that the horizontal deviation of the light caused by the schlieren diaphragm for a given vertical deviation caused in the cell depends upon the angle of the inclined slit. The smaller the angle between the slit and the vertical, the smaller is the horizontal deviation of the emergent beam. Thus rotating the schlieren diaphragm alters the proportionality between the gradient of refractive index in the cell and the z coordinate of the line on the plate. In other words the apparent magnification of the picture, in the z direction only, can be varied at will by rotating the schlieren diaphragm.

The schlieren diaphragm

In the discussion above the schlieren diaphragm was represented by a narrow slit, and indeed this was the form used originally by Svensson (1939). However the results using a slit are not very satisfactory. A photograph of a real boundary using a slit is shown in Fig. 3.8(a). The base line and the top of the peak are very thick, and there are diffraction patterns around the steep sections. The variations of thickness of the line are caused by the fact that the width of the line, measured in the z direction, is constant, being determined by the width of the slit and the magnification of the lens system in that direction. So the steep parts of the line look narrower than the flat sections. This sort of pattern is satisfactory if only the location of the top of the peak is required, but the concentration change across the boundary is proportional to the area under this peak, and this would be very difficult to measure with so much doubt as to the real position of the base line. The pattern can be improved by using a wedge-shaped slit, which produced the pattern shown in Fig. 3.8(b). This is much improved but the line still differs in thickness at different parts and the top of the peak is still thick.

Since very narrow slits are difficult to make, the alternative is to use a narrow wire on a transparent plate. This has exactly the same effect as a slit, but casts a shadow where the slit passes light and vice versa. A photograph taken using a wire is shown in Fig. 3.8(c). The base line is shown clearly but most of the rest of the peak is very poorly resolved.

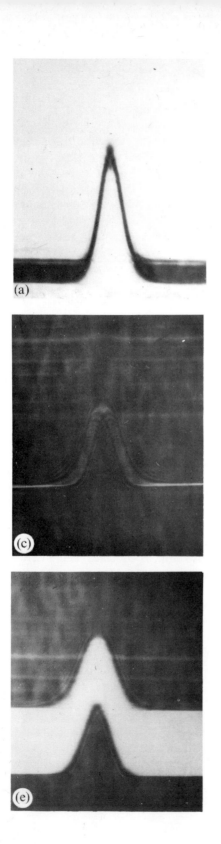

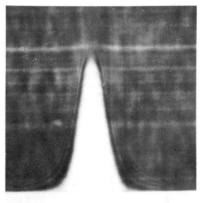

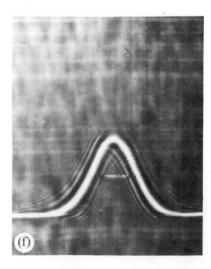

(a)

(b)

(c)

(d)

(e)

(f)

One of the most satisfactory of the early forms of schlieren diaphragms, and the form originally used by Philpot (1938), was the knife-edge which produced the picture shown in Fig. 3.8(d). This can be considered as half a slit and removes the difficulty of the varying thickness of the line produced by a slit. Diffraction effects due to the edge are inevitably present, but a more important effect was that the apparent position of the edge between light and dark parts of the picture was found to vary with the exposure given to the plate. This is largely a photographic effect and was overcome by using a bar. This may be considered as two knife-edges back to back and produced the picture shown in Fig. 3.8(e). By measuring down both edges and averaging the readings the effects of exposure and, to some extent, diffraction could be eliminated. This was the most popular of the older forms of schlieren diaphragm which were used before the phase plate was introduced.

The phase plate, introduced by Wolter in 1950, is now the universally-used schlieren diaphragm. It consists of a transparent plate half of which is covered with a layer of magnesium fluoride, or similar substance, of just such a thickness that light passing through it is impeded by exactly half a wavelength compared with light which does not pass through the layer. It produces very fine registration of the edges of the peaks but gives no image of the base line or the very top of the peak (an example of such a picture is given by Trautman and Burns (1954)). When used as a schlieren diaphragm, therefore, a very fine wire is incorporated along the edge of the magnesium-fluoride coating. This resolves the base line and the top of the peak but is so fine that it contributes nothing to the edges of the peak. The picture produced by a phase plate is shown in Fig. 3.8(f). Although this is very much clearer than those produced by the other types of schlieren diaphragm, its dual nature is revealed by a tendency for the line produced by the phase edge not to join quite accurately with that produced by the wire. This, however, is only troublesome when the most accurate measurement of the area under a peak is required.

Effects of diffraction: the phase plate

In order to understand how the phase plate works it is necessary to discuss briefly the phenomenon of diffraction, which must be described in terms of wavefronts rather than rays of light. A beam of light may be considered as an advancing wavefront which is at all points at right-angles to any individual

FIG. 3.8. Photographs of a single boundary obtained with the various types of schlieren diaphragm. (a) a slit; (b) a slit, wedge-shaped at one end; (c) a fine wire; (d) a knife-edge; (e) a bar (actually a wire about 1 mm thick); (f) a phase plate. These photographs were taken by the author under circumstances in which it was very difficult to ensure that the angle of inclination was exactly the same for each diaphragm; consequently the heights of the peaks are not exactly the same, as they should be if this angle had been kept constant.

ray. Thus a parallel beam of light has a plane wavefront, a beam diverging from a point has an expanding spherical wavefront, and a beam converging towards a point has a contracting spherical wavefront. The progress of a wavefront can be described by assuming that each point on the wavefront consists of a new source of light, and the combination of the spherical wavefronts from these points forms the new wavefront. Interference effects (see Chapter 4) result in the original wavefront moving forward at right-angles to itself at all points, so long as the edge of the beam is remote from the point of interest. However, if an edge is inserted into the beam, the emerging wavefront cannot end abruptly and the light appears to bend round the edge so that the edge of the shadow is not sharp. This is illustrated in Fig. 3.9; Fig. 3.10 shows the light intensity across the edge of the geometrical shadow.

The important point to notice in Fig. 3.10 is that the geometrical edge to the shadow (as determined by drawing straight lines in the direction of the beam) is not at the point where the illumination is half that remote from the edge but at the point where the illumination has fallen to one quarter. Also

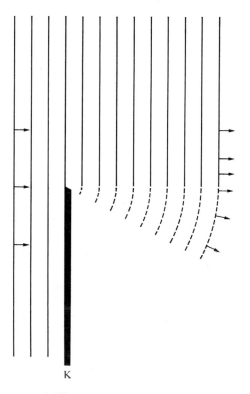

K

FIG. 3.9. The phenomenon of diffraction. A plane wavefront is shown approaching a knife-edge K from the left. After passing K the wavefront appears to spill around the knife-edge and moves forward as indicated by the arrows.

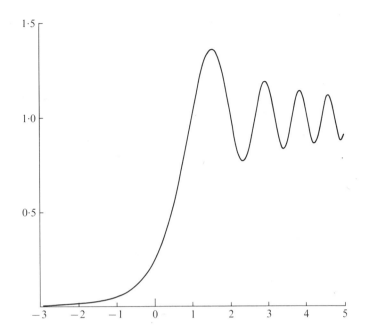

FIG. 3.10. The variation of the intensity of the light across the geometrical edge of the shadow of a knife-edge illuminated by pure monochromatic light. The geometrical edge of the shadow is at zero on the x-axis. In practice the pattern is seldom so well defined because no light source is perfectly monochromatic. In the context of the optical systems described in this book, only two or three of the maxima in the intensity are observed and the ratio of the maxima to the minima are reduced. The curve was obtained by numerical integration of the Fresnel integrals, and for a full description of these the reader is referred to almost any advanced text book on optics.

the light level in the illuminated sector varies over quite large distances in a regular way, and this effect gives rise to what are known as diffraction fringes. Note also that the light level at first increases to a level about 37 per cent above the level that would have been observed in the absence of the opaque edge. This point of maximum intensity or the point of minimum intensity following it are the easiest points to locate, but they are displaced appreciably from the geometrical edge of the shadow. When a photograph is taken of this shadow, the nonlinear response of the film can result in the apparent edge being almost anywhere along the first steeply rising part of the curve, depending on the exposure given. This was the reason earlier for using a bar instead of a knife-edge for the schlieren diaphragm.

It should be further pointed out that the diffraction effects illustrated above are only obtained in planes where the edge is not in focus. Obviously no such effect is observed in the plane of the edge itself; nor is it observed if an image of the knife-edge is formed by a lens. When it comes to applying these concepts to the schlieren optical system, it is important to realize that the astigmatic camera focuses the knife-edge on to the plate in the z direction only. Thus when the image of the edge is horizontal (on the plate) it is in focus and no diffraction effects are observed. On the edges of the peak, however, it is no longer in focus and the diffraction effects are seen (see Fig. 3.8). This applies equally to the knife-edge or the edges of a slit or a wire.

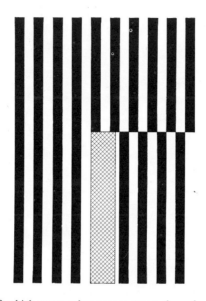

FIG 3.11. The phase shift which occurs when a wave passes through a phase plate. The coating of the phase plate is indicated by the hatched area, and the black and white stripes represent a wavefront moving from left to right. The part of the wavefront which passes through the coating is impeded in such a way that it emerges just half a wavelength out of phase with the part of the wavefront which did not pass through the coating. It can be seen that in the direction of the geometrical edge of the shadow of the coating there is no alternation of black and white, and this represents zero intensity of the light. Diffraction effects spread each part of the wavefront into the other so that the effect is not as sharp as implied in the diagram.

Consider now a plane wavefront approaching a phase plate (Fig. 3.11— in which the glass supporting plate has been omitted for clarity). The edge of the plate divides the wavefront into two parts, one of which passes through the coating and one of which does not. In the absence of the coating these two wavefronts would, of course, recombine to give another plane wavefront which would illuminate a screen evenly. Each half would produce a diffraction pattern but these would cancel one another out. However, the phase coating causes the half of the wavefront passing through it to emerge exactly half a wavelength out of phase with that part which did not pass through it. Consequently, when the two diffraction patterns recombine, they cancel where they would normally reinforce and they reinforce where they

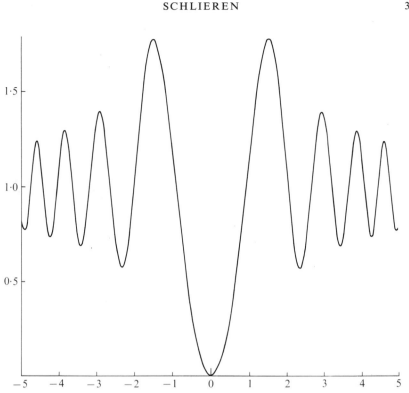

FIG. 3.12. The variation of the intensity of the light across the geometrical shadow of a phase edge illuminated with pure monochromatic light. The curve was calculated by summing the effects of two diffraction patterns, as shown in Fig. 3.10, allowing for the difference in phase between the two parts of the wavefront. As with Fig. 3.10 the effect is never as clear as indicated here, but the intensity of light at the edge of the geometrical shadow is zero, or very nearly so.

would normally cancel. The result is an exaggerated diffraction pattern which has zero intensity at the geometrical edge of the shadow (Fig. 3.12). The spacing of the diffraction fringes is the same for the phase plate as for a knife-edge, but the ratio of the intensities of the light at the first maximum and the following minimum is larger for the phase plate than for the knife-edge. The theoretical values are 3·1 for the phase plate and 1·7 for the edge, but neither value seems to be achieved in practice. This may be due to photographic effects or to imperfections in the optical system, but it is observed that the fringes are clearer in a phase-plate picture than in a knife-edge picture. So the phase plate not only produces zero light intensity at the geometrical edge of the shadow, but also produces clearer diffraction fringes around this minimum, which are useful for accurate location of the minimum. Also, since the centre of the pattern is a true zero of intensity, large photographic exposures can be used which increase the effective contrast and again ease the location of the point of zero intensity.

It is because the phase plate produces its pattern by the use of the phenomenon of diffraction that it can only produce a line in planes where the edge is
not in focus. Thus by itself it cannot show a base line or the top of a peak
where the edge is in focus on the plate and no diffraction effects are seen with
any sort of schlieren diaphragm (see Fig. 3.8). Ways of getting around this
difficulty include defocusing the optics, putting the phase edge at the centre
of a narrow slit, or incorporating a fine wire along the phase edge. Of these
the last is the most generally adopted since it works well and does not interfere
with the system when used as an interferometer (see Chapter 4). One manufacturer has incorporated a second thin wire set at an angle of 25° to the
phase edge. The result is that at one particular setting of the phase-plate angle
the extra wire is at right-angles to the images of the light source and results
in a base line throughout the cell. This has great advantages when the area
under the peak is required, but it works only at the one particular angle. This
is less of a disadvantage in practice than it might seem in theory, because the
clarity of the line decreases as the angle is increased, so that there is a
compromise between a sharp line and a large deflection, and in practice only
a small range of angles is normally used.

The Centriscan system

The scanning absorption optical system of the M.S.E. Centriscan described in Chapter 2 (p. 13) may be modified to measure gradients of refractive index. An extra correcting lens and an interference filter for 550 nm are
introduced just above the rotating rotor (see Fig. 2.5). The extra lens produces an image of the mirror slit just in front of the photomultiplier. At this
image point is placed a knife-edge set at an angle to the mirror slit. Thus the
mirror slit, correcting lens, and knife-edge correspond to the first part of the
schlieren system up to and including the schlieren diaphragm. As the
scanning slit moves the beam of light through a gradient of refractive index,
the image of the mirror slit in the plane of the knife-edge moves so that the
amount of light intercepted by the knife-edge varies. The system is so
arranged that the refraction of light in the cell results in less light passing the
knife-edge so that the photomultiplier sees an increased absorption in the
system. Thus a plot of the absorption of the system as a function of the
position in the cell (which is what the absorption system records) is also a
plot of the gradient of refractive index as a function of the position in the
cell (which is what the schlieren system records).

Results

The schlieren optical system produces a pattern which is a direct plot of
the gradient of refractive index as a function of the distance down the cell.
Since refractive index may normally be taken as a linear function of the
concentration, the plot becomes one of the concentration gradient dc/dx as

a function of x. Since $\int (dc/dx)\,dx = \Delta c$, the area under the schlieren line represents the change of concentration between the two limits of integration. Thus the patterns may be used to evaluate the composition of a mixture if the components are sufficiently well resolved.

The advantages of the system are that it uses visible light and produces patterns which are readily evaluated by eye. The differential nature of the measurement reveals overlapping boundaries more readily than the corresponding integral curve would do (see Fig. 2.4). If the integrated form is needed the area under the curve yields this information without much difficulty.

The disadvantages are the indiscriminate nature of a method depending on refractive index, a comparative lack of sensitivity compared with the absorption system with a strongly absorbing solute, and a lack of precision. This latter arises from the width of the line produced, even by the phase plate. This can be improved to some extent by heavy exposure of the photographs, but this makes the measurements rather more difficult and, in any case, is not always possible. Since the phase plate uses the phenomenon of diffraction itself to produce the line, it may well represent the best that the wave nature of light will permit.

4

INTERFERENCE

The phenomenon of interference

A BEAM of light is an oscillating electric field which is propagated through space at a finite velocity. (Associated with this electric field is an oscillating magnetic field, but this need not concern us here.) For a pure monochromatic beam in one dimension the variations of the electric field strength may be represented by the simple sine curve shown at the top of Fig. 4.1, in which the horizontal axis may be either time or distance. If two beams of the same frequency and amplitude are considered, these may differ in phase. The two extreme cases are shown in Fig. 4.1. At *a* the two waves are exactly in phase with one another, and at *b* they are exactly out of phase; (the two waves at *b* are said to have a phase difference of π radians or 180°). Since the electric field oscillates between equal positive and negative amplitudes, if the two beams are combined, in *a* they add together to produce a wave of twice the original amplitude, whereas in *b* they add to give zero amplitude since the positive excursions of one wave cancel exactly the negative excursions of the other. These are, of course, extreme cases. In general the phase difference may be anything from 0° to 360° and the amplitudes of the two waves may differ. In general, therefore, when the waves are added, a new wave is produced with the same frequency as the component waves but of an amplitude which may have any value between zero and the sum of the two individual amplitudes depending on the relative phase and amplitudes of the constituent waves.

(a)

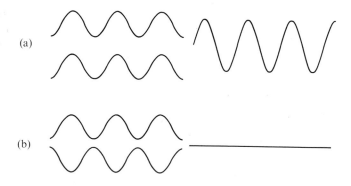

(b)

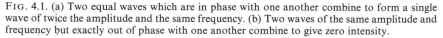

FIG. 4.1. (a) Two equal waves which are in phase with one another combine to form a single wave of twice the amplitude and the same frequency. (b) Two waves of the same amplitude and frequency but exactly out of phase with one another combine to give zero intensity.

For the phenomenon of interference, as described above, to be observed in practice an important condition has to be satisfied. The two beams of light must be *coherent*, which means that the two beams must have the same phase with respect to one another over the whole of the time covered by the observations. Beams from two different light sources, however similar, are never sufficiently coherent because the beam from a light source (even a laser) does not consist of a single continuous train of waves but of a large number of short wave trains which have arbitrary phase relationships with respect to one another. Thus if the beams from two different light sources are combined, although interference occurs, the positions of the maxima and minima fluctuate rapidly as the phase between the two light beams changes so that interference patterns cannot be observed. The only way of obtaining two beams with sufficient coherence is to derive them both from a single original beam. Basically there are two different ways of doing this.

The first system of beam splitting is illustrated in Fig. 4.2 and is the one

S_1 S_2

FIG. 4.2. The method of obtaining two beams with a constant phase difference by means of Young's slits. A single slit S_1 is illuminated by light from a source not shown, and light from this slit spreads by diffraction to illuminate two slits S_2 from which the two beams with constant phase difference emerge.

F IG. 4.3. The combination of two beams of light to form interference fringes. The diagram shows the two slits S_2 of Fig. 4.2 illuminated from a single slit very far to the left. The wave is represented by the alternating black and white lines and these emerge as arcs of circles centred on the two slits. It can be seen how the combination of the two beams produces beams of light interspersed by dark areas. Light, in this diagram, is represented by alternating black and white lines.

used for the first demonstration of true interference by Young. The narrow slit S_1 acts as a light source which illuminates a second screen S_2, which has two narrow slits parallel to the first. The two slits in S_2 act again as light sources but since they derive their energy from the waves coming from S_1, so long as the distances from S_1 to the two slits in S_2 are about the same, there will be a constant phase difference between the beams from the two slits in S_2. If the slits in S_2 are very narrow, the beams from them will spread out, owing to diffraction, and so overlap to the right of S_2. The interference effect can be visualized as shown in Fig. 4.3, in which the screen S_2 of Fig. 4.2 is shown. The source slit is assumed to be so far to the left that the wavefronts approaching S_2 are plane. The positions of the wavefronts have been drawn at intervals of a wavelength and it can be seen how the light emerging from the two slits appears to form a series of beams which form the interference pattern on a screen to the right of the diagram. In practice the two slits are usually too wide and too far apart for this simple arrangement to work because the beams emerging from the slits do not spread out sufficiently to overlap, so a lens is placed to the right of the slits and this focuses an image of the source slit where the interference lines are seen. This is the basis of the most commonly used form of interferometer, the Rayleigh interferometer.

The other way of splitting a beam into two components is to use a partially reflecting surface. At such a surface part of the beam passes through the surface and part is reflected forming two beams which are coherent. The partially reflecting surface may be made the front surface of a block of glass with the back surface totally reflecting. This results in the beam transmitted by the first surface being reflected by the second so that the two beams emerge parallel to one another (Fig. 4.4). A similar plate is used to recombine the two beams. This is the basis of the Jamin interferometer.

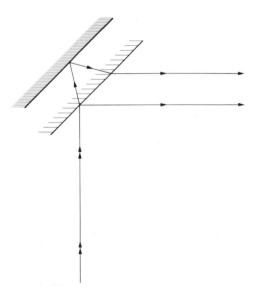

FIG. 4.4. The method of producing two beams of constant phase difference which is used in the Jamin interferometer. The first surface is partially reflecting and the second totally reflecting. This arrangement results in the two beams emerging parallel to one another.

One other form of interferometer has been used for free-boundary diffusion measurements. This is the Gouy interferometer, and beam splitting is achieved by the boundary itself. A brief description is given at the end of this chapter.

The Rayleigh interferometer
The optical system

As with all forms of interferometer (except the Gouy) a double cell is used, one half of which is filled with the solution under study and the other half of which is filled with a reference solution (usually the solvent in which the macromolecule is dissolved). The two coherent beams of light are passed one through each half of the cell and they are then recombined. A horizontal cross-section of the Rayleigh interferometer is shown in Fig. 4.5, and the similarity to Fig. 3.6 is obvious. L_4 is a cylindrical lens for the same reason as in the Schlieren system (see p. 26). (This system is a modification by Svensson (1950a) of a system described by Philpot and Cook (1947) and has been referred to as the Rayleigh–Philpot–Cook–Svensson system.) It is in fact identical with the schlieren system except that the source slit is rotated through 90° and shortened, a double cell is used, and a double slit is interposed as a beam splitter. The schlieren diaphragm is not used but, unless it is a slit, it need not be physically removed; rotation to the vertical position effectively removes it from the system. Light from the source is rendered parallel by the collimating lens L_1, and passes through the cells and the double slit (which may be anywhere between L_1 and L_2). The condensing lens L_2 produces an image of the source slit at its focal point K. At this point the interference lines are produced, the positions of which depend on the difference in optical path length between the source and K down the two channels. If the solutions in the two cells are uniform throughout, the interference lines at K will be straight and visible. If, as is more usual, one cell is full of a uniform solution and the other contains a solution whose refractive index changes with position in the cell, the image at K is made up

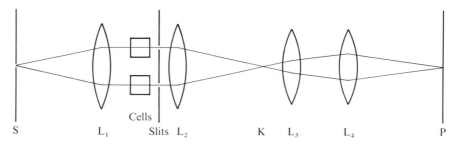

S L_1 Slits L_2 K L_3 L_4 P

Fɪɢ. 4.5. The Rayleigh interferometer in horizontal cross-section (the *yz* plane). L_4 is a cylindrical lens with its axis vertical and so appears like a normal lens in horizontal cross-section.

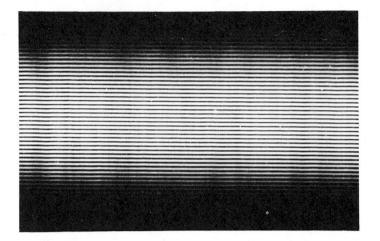

FIG. 4.6. A typical photograph of the interference fringes produced by the Rayleigh interfero meter with a single-slit source and interference slits about 0·1 mm wide.

of a large number of sets of interference lines at slightly different positions horizontally so that no interference lines are visible at K. The astigmatic camera consisting of spherical lens L_3 and cylindrical lens L_4 focuses the cell in vertical coordinates and the interference pattern at K in horizontal coordinates onto the screen or photographic plate P.† The resulting pattern when both cells are filled with a solution of uniform refractive index is shown in Fig. 4.6. The convention with these patterns is that the top of the cell is on the left, the bottom of the cell on the right, and an increase of refractive index causes the fringes to move upwards. The separation of the fringes is determined by the separation of the two slits, the focal length of lens L_2, the magnification of the astigmatic camera in the z direction,‡ and the wavelength of light. The width of the band of light in which the fringes are seen (the diffraction envelope) is determined by the Fraunhofer diffraction at the slits. The narrower the slits the wider is the diffraction envelope and so the greater is the number of interference lines within it. However, not only do narrow slits pass less light, but this light is spread over a larger area of image, so that the intensity of the illumination of the image decreases rapidly as the slits are narrowed.

A way of increasing the number of interference lines which appear on the plate without reducing their separation has been described by Svensson (1951) and is available commercially. The light source instead of being a single slit, as described above, is a grating consisting of a very large number

† For a more detailed description of the astigmatic camera see Chapter 3.

‡ For the definition of the x, y, z directions see p. 4.

of slits side by side. Each slit acts as a source and produces a set of interference fringes, and the spacing of the slits in the grating is adjusted so that
the sets of fringes produced by adjacent slits are displaced by exactly the
separation of the fringes. Thus all the fringes match together and the result
is, perhaps, a hundred or so fringes instead of the nine or ten produced by a
single slit. This has several important advantages over a single slit, not the
least of which is that the fringes so produced are all of equal intensity and
very much brighter than those from a single source slit. Not only does more
light come through a hundred slits than through one slit, but the two slits in
the beam splitter do not need to be so narrow, because the system does not
depend upon diffraction at these slits to produce a diffraction envelope in
which the fringes are seen. A further advantage is that it becomes possible to
use a lower power lamp to illuminate the source which in turn means that
the pressure within a conventional gas-discharge lamp can be lower and this
reduces the bandwidth of the emitted line. This improves the visibility of the
fringes when there is a relatively large path difference between the two
channels.

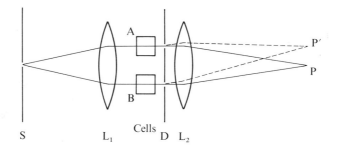

FIG. 4.7. The first half of the Rayleigh interferometer system illustrating the movement of
fringes when the refractive index in cell A increases. For details see the text.

The effect of changes in the refractive index of one of the solutions in the
double cell can be described with the aid of Fig. 4.7, which shows the optical
system, omitting the astigmatic camera, in horizontal cross-section. The
solid lines represent the rays of light which give rise to a particular fringe at
P. For the sake of argument we will assume that this fringe represents zero
difference in path length from S to P via the two cells A and B when the
refractive indices of the solutions in the two cells are the same. Suppose now
the refractive index of the solution in cell A increases slightly, then the
optical path length through cell A increases (see Chapter 1) so that point P
will no longer be at equal optical distance from S via both channels. However, at some point P' the increased physical distance from the lower slit in
the diaphragm D, combined with the decreased distance from the upper slit,

will compensate for the increased optical path through cell A. Thus the fringe originally at P will appear to move to P'. This will be visible so long as P' lies within the area illuminated by diffraction effects at the slits in D.

Results

The image produced by the astigmatic camera consists of the interference pattern produced by a pair of rays passing through a given level in the cell as a function of that level. Thus, as the refractive index of the solution increases down the cell, the fringes move across, and the result (rotated by convention through 90°) for a single-slit source, is shown in Fig. 4.8. The movement along the image of Fig. 4.8 from one fringe to the next represents a change in optical path length from the source to the plate of one wavelength of light. Thus, if the movement of one fringe across the boundary, represented in Fig. 4.8, is measured in terms of the separation of the fringes, the result is the difference in the number of wavelengths contained within the solution cell (cell A in Fig. 4.7) between a level above the boundary and below it, less any change which may have occurred between these levels in the reference cell (B in Fig. 4.7). Since the optical path length in the cell is given by na/λ, where n is the refractive index, a is the physical path length, and λ is the wavelength of light in vacuo, this difference is $a\Delta n/\lambda$ where Δn is the change in refractive index across the boundary. As was shown in Chapter 1, Δn is proportional to the concentration change, so the lines in Fig. 4.8 are plots of the concentration of the solution as a function of the distance down the cell. (This assumes that the concentration of the reference solution is constant, but it must be remembered that the system measures the difference in concentration between the two halves of the cell so that any changes in one cell are subtracted from the other automatically; this is one of the very useful aspects of the system, especially in the ultracentrifuge, where sedimentation of buffer components can cause appreciable changes in the refractive index of the solutions.)

Although the lines of the interferograms are plots of the concentration of the solution as a function of the distance down the cell, they give no indication of the position of the origin of the plot. In the example shown in Fig. 4.8 it might reasonably be assumed that the area to the left of the boundary is

FIG. 4.8. The interference pattern produced by a single boundary between a solution and its solvent when the reference cell contained a solution of uniform refractive index. The optical system was similar to that used to obtain the photograph shown in Fig. 4.6, but the interference slits were about 0.3 mm wide.

zero concentration. It is indeed zero concentration of the migrating species, but the picture gives no indication of whether the refractive index of the solution in that region is the same as that of the reference solution or not. For migration experiments this is not important, but in sedimentation-equilibrium experiments it is a serious drawback. Although several methods have been derived to overcome this drawback, only one of them is strictly an optical method. The inability of the interference system to indicate the absolute difference in refractive index between the two solutions arises because, when monochromatic light is used (as it usually is), all fringes look exactly the same. So a uniform increase in refractive index throughout a solution simply moves the whole pattern across and the new pattern is indistinguishable from the old. However, this is not true if white light is used instead of monochromatic light. As mentioned above, the separation of the fringes in the interferogram depends, among other things, on the wavelength of light. With white light, therefore, no fringes are generally visible, because those in various colours have different separations and combine to give a general blur in the photograph (at least in a monochrome photograph). This is true in all places except one, and that is the point where the optical path lengths from source to plate down the two channels are exactly the same. At this point all colours will arrive in phase, and a bright line results. Since in practice white light consists of only a relatively narrow band of wavelengths (covering only about a factor of two in wavelength), as one moves away from the bright central fringe all waves present begin to go out of phase together and so the central bright fringe is bordered on each side by a dark fringe. In practice, about three bright fringes are visible. A picture of this is shown in Fig. 4.9, in which it will be noted that on either side of the central clear fringes are other sets of fringes which are poorly resolved. This arose because the source of light was a mercury-arc lamp and this emits most of its light at the two wavelengths 546 nm and 436 nm, which happen to be very nearly in the ratio 5:4, so that after 4 green fringes and 5 blue fringes the two sets of fringes nearly match again. There is generally no difficulty distinguishing these from the true achromatic (or white-light) fringe, and indeed they can sometimes be useful. If the optical system is set up initially so that the achromatic fringe is in the centre of the diffraction envelope when the solution in the two sides of the cell are of equal refractive index, and the cells are then filled with solutions of uniform but different refractive index, this difference of refractive index can be measured by the displacement of the achromatic fringe away from its original position.

In order to observe the achromatic fringe, the optical system must be set up in such a way that the optical distances from the light source to the centre of the photographic plate, through the two paths, are the same to within a wavelength of light. If this is done with no cell in position, and then a cell containing solvent in one half and a solution of higher refractive index in the

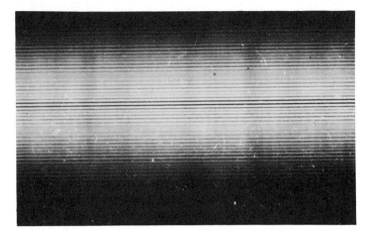

FIG. 4.9. The achromatic fringe.

other is inserted, the equality of optical path length is destroyed and the achromatic fringe will generally move out of the diffraction envelope. There are two possible solutions to this problem. If the layout of the optical system permits, a second double cell containing the same solutions, but the opposite way round, may be introduced to compensate for the difference in path length in the experimental cell. This is undoubtedly the best solution, but unfortunately most optical systems are not designed so as to make this possible. The alternative is to add to the reference solvent a solute which will raise its refractive index to that of the solution. For sedimentation work butane-1:3-diol has been recommended (Richards and Schachman 1959) and in the author's experience this works well.

The Rayleigh interferometer has the advantages over the schlieren system of increased sensitivity and accuracy. For a typical protein ($dn/dc = 0.183$ ml g^{-1}) a movement of the fringes by one fringe spacing corresponds to a concentration change of about 3×10^{-4} g ml^{-1} in a cell 1 cm long, and this figure is independent of magnifications within the system, and so applies to any system working on these principles. The movements of fringes can usually be measured to better than one tenth of a fringe spacing so an accuracy of about $\pm 3 \times 10^{-5}$ g ml^{-1} can be attained easily. It suffers from the same disadvantage as any refractometric method in that it cannot distinguish one solute from another, but the system automatically subtracts the base line from the pattern. If the reference cell contains not a simple solvent but a solution containing migrating species the system will measure directly the differences between the two solutions. The commonest use of this kind of facility is in sedimentation-equilibrium experiments, where the small concentration gradients caused by the sedimentation of the buffer

salts are automatically subtracted from the gradients in the solution cell and
the optical system responds only to the changes in concentration of the
macromolecule. It has also been used to determine the difference between
the sedimentation coefficients of two very similar proteins (Richards and
Schachman 1959; Kirschner and Schachman 1972). The output is an
integral curve which shows overlapping boundaries less clearly than a
differential curve (Fig. 2.9).

Combined schlieren and interference records

The great similarity between the schlieren optical system (see Chapter 3)
and the Rayleigh interferometer (cf. Figs. 3.6 and 4.5) raises the possibility
of combining the two to produce simultaneous records of the integral
(n versus x) and derivative (dn/dx versus x) curves. The most important
difference between the two systems is in the direction of the source slit, which
is vertical for the interferometer and horizontal for the schlieren system.
Superposition of these two slits produces the desired result, but the interference record is largely obliterated by the light from the schlieren slit. The
two images can be separated by separating the source slits, but one or other
must then be off-axis. A more elegant solution is to use a multi-slit interference source and to mask it with the schlieren slit. The source then consists
of a row of point sources. This arrangement was suggested by Svensson
(1951), who also shows the results of using this with all possible schlieren
diaphragms except the phase plate, which had only been described the year
before. The results are more or less as might be expected. With a knife-edge
as schlieren diaphragm the fringes are obliterated below the derivative contour, and with a wire they are obliterated within the derivative contour. A
simple slit could not be used as it intercepted the reference beam required by
the interference system. Instead a double slit was used with one of the slits
fixed parallel to the source slit to pass the reference beam. Interference
fringes were obtained within the derivative contour only, except where the
gradient of refractive index was zero, where interference fringes were
obtained right across the plate.

Differential interference

It is possible to obtain derivative records (dn/dx versus x) directly from
interference optics. A single cell is used and the two interfering beams are
arranged to pass through the solution at very slightly different levels. Two
ways of doing this have been published. The first was by Svensson (1950b;
Svensson, Forsberg, and Lindstrom 1953) in which he introduced four
additional parallel-sided blocks of glass into the optical system (Fig. 4.10).
Two of these were placed side by side just before the cell and the others
were placed just after the cell. The blocks were placed vertically but inclined
at a small angle along the optical axis, one of each pair leaning towards the

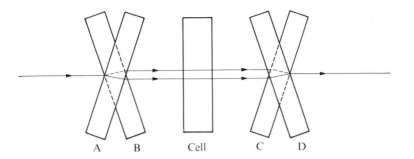

FIG. 4.10. One method of using interference optics to produce derivative (dn/dx versus x) curves directly. A,B and C,D are parallel-sided blocks of glass placed side by side close to the cell. A narrow sheet of light, entering the system from the left, is split by blocks A and B into two parts which pass through the cell at slightly different levels, and these are recombined by blocks C and D.

cell and the other away from it. The blocks on the other side of the cell lent in the opposite directions. The effect of these blocks of glass was to split each part of the incident beam into two, one raised slightly and the other lowered slightly with respect to the original level. These two beams passed through the solution and were combined again by the blocks of glass on the exit side of the cell. Thus the two interfering beams passed through the cell at slightly different levels, and so experienced different path lengths to an extent depending on the value of dn/dx at that level. The second method was published by Wiedemann in 1952. The principle was just the same but, instead of two blocks of glass, he used a prism similar to a Fresnel prism (apex angle close to 180°), with its base normal to the optical axis and its apex edge parallel to the cell. This, together with a source slit inclined slightly to the vertical, produced two beams passing one over the other in the centre of the cell. A second prism, together with the cylindrical lens tilted at the same angle as the source slit but in the opposite direction, recombined the beams to produce the derivative record on the photographic plate.

Both authors claim that the system is more accurate than the schlieren system, but neither has been taken up commercially, and, as far as this author is aware, no instance of its use appears in the literature.

The Jamin interferometer

As with the Rayleigh interferometer, two cells are used, one containing the solution under study and the other a suitable reference solution. Two coherent beams that traverse these cells are produced by a beam splitter of the type shown in Fig. 4.4. The first surface is partially reflecting, and should ideally reflect about 38 per cent of the light falling on it, and the second surface is totally reflecting. The two beams, after traversing the two cells, are recombined by a second, similar, pair of surfaces. The fringes so produced are localized at infinity if all the reflecting surfaces are parallel. However, if the second pair of surfaces are inclined very slightly to one another, the fringes may be localized in the cell and a single lens will then focus them and the cell coordinates simultaneously onto a photographic plate. The system resolves both vertical and horizontal coordinates in the cell and, although this is not strictly necessary, it may be useful for detecting and allowing for slight inhomogeneities in the cell windows.

A discussion of various forms of this interferometer applied to free-boundary electrophoresis experiments has been published by Lotmar (1947), in which he describes some interesting variants of the system, in which the light is passed twice through the cell and one pair of reflecting surfaces is used for both splitting and recombining the beam. The use of the system for ultracentrifuge work has been described by Beams, Snidow, Robeson, and Dixon (1954). The optical system used by these authors is somewhat complex and the reader is referred for details to the original paper or to the book by Schachman (1957) where the optical diagram is reproduced. As used by them provision is made for the use of a compensating cell so that the achromatic fringe can be observed.

The Gouy interferometer

This form of interferometer is applicable only to boundary systems for which the concentration-gradient curve is symmetrical about a single maximum (Gosting 1956). Within these restrictions it is both more accurate and quicker than other optical means of studying free diffusion (Coulson, Cox, Ogston, and Philpot 1947). A beam of light from a horizontal slit is made parallel by a collimating lens and illuminates a single cell in which the diffusing boundary is set up. A condensing lens following the cell pro-duces an image of the horizontal slit at its focal point. (In some versions of the interferometer a single lens combines the functions of the two lenses described above, but in this case the cell is not illuminated with parallel light.) Rays of light which pass through the boundary layers are refracted towards the side of high refractive index and, when brought to focus by the condensing lens, rays which pass through adjacent layers interfere with one another to produce a series of interference fringes. This effect was first described by Gouy in 1880, and the theory of the method as applied to the measurement of free diffusion was first described by Coulson et al. (1947) and by Kegeles and Gosting (1947), and further developed by Ogston (1949) and by Gosting and Onsager (1952). The theory has been verified by Coulson et al. (1947) and by Longsworth (1947), and has been used extensively for the measure-ment of the diffusion of small molecules and salts both in binary and ternary systems (Gosting and Morris 1949; Gosting and Akeley 1952; Akeley and Gosting 1953; Dunlop and Gosting 1959, 1964; Fujita and Gosting 1960).

5

ALIGNMENT OF THE OPTICAL SYSTEMS

I F any optical system is to yield patterns of the highest quality, it is necessary that all the components are accurately aligned on a common optical axis and positioned to yield the best focus of the image onto the photographic plate. The exact procedures necessary to achieve this alignment will vary slightly according to the detailed design and layout of the individual system, but in this chapter the principles involved will be discussed together with a more detailed indication of the methods which may be used to align a combined schlieren and Rayleigh-interferometer system. The alignment of an absorption system follows exactly similar lines but certain steps which are unique to the schlieren or interference system are not needed. A detailed description of the alignment of one commercial system has been given by Gropper (1964), and the methods given here are largely based on his procedure. More recently a more detailed description of the alignment of this same system has been given (Richards, Teller, Hoagland, Haschemeyer, and Schachman 1971; Richards, Teller, and Schachman 1971; Richards, Bell-Clark, Kirschner, Rosenthal, and Schachman 1972).

It will be assumed that all the components shown in Figs. 3.6 and 4.5 are present, together with one or two plane mirrors to bend the optical system, since most commercial systems are bent in this way. It will be further assumed that all the optical components can be adjusted in the three dimensions in space and rotated about three mutually perpendicular axes. These axes will be taken as those designated x, y, z, in Fig. 1.2, of which y is the optical axis. This means that each component should be capable of adjustment in six ways, but in practice only five are ever needed, because spherical lenses do not require rotational adjustment about the optical axis and the cylindrical lens needs only a general alignment along its own axis. Not all commercial systems allow for the full adjustment of all the components and, indeed, this is not always necessary because fixed components can be used to define the optical axis. When plane mirrors are included, they may be adjusted to cause the optical axis to pass through other fixed components.

In the discussion which follows reference to the light source will refer to the horizontal or vertical slit that is the effective source for the schlieren or interference system. The lamp which supplies light to this source will, in general, be situated behind the slit, and the light from it may or may not be focused onto the slit by a lens. During the preliminary setting up it will be necessary to place this lamp also on the optical axis, but such adjustment is not very critical and, so long as the light from the lamp passes through the source slit and illuminates the collimating lens evenly, the position of the lamp is satisfactory.

Preliminary setting up

The first requirement is to define the optical axis from the light source to the cell. In a static system, if all components are fully adjustable, the selection of the components to define this axis is somewhat arbitrary. However, in practice, especially in the ultracentrifuge, the cell is likely to be the least adjustable component, and it is the natural component to select to define the optical axis. This axis will be defined here as the line that passes through the centre of the cell and is normal to the entrance and exit windows. Some workers with an ultracentrifuge have used the earth's gravitational field to define the optical axis through the cell. This is satisfactory because the rotor hangs freely in the earth's field. However the same workers use the earth's field also to define the x axis at the photographic plate. Since the x axis is at right-angles to the optical axis at the cell, this rotation of 90° must depend upon the accuracy with which the optical bench is set at right-angles to the radial direction at the cell. It applies only to the ultracentrifuge (and possibly to only one make of ultracentrifuge) and is inapplicable to any other apparatus. The author cannot recommend it as a general procedure, although it may be satisfactory in a particular case.

First, it is necessary to place the light source and the collimating lens (L_1 in Fig. 3.6) in position on the optical axis and to place the light-source slit at the first focal point of the collimating lens. If the focal length of the collimating lens is known, this last adjustment is a simple matter of measurement. If the focal length is not known it may be determined accurately enough for this purpose by using the lens to project an image of a distant object onto a screen, when the focal length is the distance from the lens to the screen. A more precise adjustment is made later. The collimating lens is normally placed as close as possible to the cell. Having placed the light source near the focal point of the collimating lens, an approximately parallel beam of light will be projected towards the cell. If the line joining the light source to the centre of the collimating lens is parallel to the optical axis, light reflected by the windows of the cell will return through the collimating lens and be brought to focus at the light source. The amount of light reflected by a glass window is small (about 4 per cent at each surface), so that this adjustment is difficult unless one window is treated to make it totally reflecting. This may be done by coating one window with aluminium by evaporation in a vacuum, or by deposition of silver from an ammoniacal solution of silver nitrate containing a mild reducing agent such as glucose. The former is more durable and the latter may be removed very easily by treatment with dilute nitric acid. If the light source is not exactly in the focal point of the collimating lens, the image formed by the reflected light may be in front of or behind the source. The light source may be adjusted along the optical axis until the source and the image are in the same xz plane.

Having ensured that the line joining the centre of the light source to the centre of the collimating lens is parallel to the optical axis, it is necessary to ensure that it is coincident with it. To do this the size of the light source is reduced as much as possible and centred at the centre of the normal slit, and all but the centre of the collimating lens is masked off with an opaque screen. This combination produces a narrow beam of light which should strike the centre of the cell. The lens and light source are adjusted in the x and z directions to achieve this. The former adjustments should then be rechecked to ensure that the lens and source have been moved an equal amount. Finally the source can be placed accurately at the focal point of the lens as described below. It may be necessary to adjust the position of the lamp at this stage.

If there is a plane mirror between the light source and the cell, it is possible to adjust this in place of either the lens or the light source. However, it must be emphasized that, once a position of such a mirror has been selected, it should not be readjusted later as this can affect the rotational adjustment of the light source.

Having placed the light source and the collimating lens on the optical axis, as defined by the cell, it is necessary to extend this line through the cell as far as the photographic plate, and to place all other components on this axis. If the small source and the mask on the collimating lens that were used in the last step are retained, but the reflecting coating on the cell is removed, a thin beam of light will pass right through the optical system. This defines the optical axis. If a plane mirror is incorporated between the cell and the photographic plate, it should be adjusted first to cause the beam of light to strike the centre of the plate. Alternatively the camera is adjusted to achieve the same result. At this point it is essential to be able to replace the photographic plate by a ground-glass screen in order to observe the images directly. This screen should be marked with a vertical and horizontal line passing through the centre of the screen. The beam of light should strike the screen at the intersection of these two lines. Again it is important that, once a position has been found for any plane mirror in this part of the system at this stage, it must not be altered later because, although movement in one direction will simply cause the image on the screen to move in that direction, movement in the other direction will cause the image to rotate and so upset the rotational adjustments of the light source, masks, and the cylindrical lens. If any of the components between the cell and the camera are not fully adjustable, it may be necessary to move the mirror along the y axis in order to raise or lower the optical axis with respect to the optical bench to which the components are attached. This will be necessary, for instance, if the schlieren diaphragm cannot be moved in the z direction. In any case it is best to ensure that the optical axis is parallel to the optical bench.

The rest of the components can now be placed in position one by one. There is some advantage at this stage if the mask over the collimating lens is replaced by one serving the same purpose at the centre of the cell. This mask can then be used for setting the approximate position of the camera lens by observing the image of this mask on the ground-glass screen. The lenses are placed in position in the following order: condensing lens L_2 (Fig. 3.6), camera lens L_3, cylindrical lens L_4. After each lens is placed in position, it is adjusted in the x and z directions so that the patch of light on the screen remains at the point of intersection of the two lines drawn on the screen. It is then adjusted for rotation about the x and z axes, so that the small amount of light reflected from the lens passes back along the optical axis. The cylindrical lens must be put in with its axis parallel to the x-axis of the cell and adjusted in this direction so that the light passes through the middle of the lens. This lens will cause the image on the ground-glass screen to spread into a line in the z direction, but this line should stay on one of the lines on the screen and be symmetrically placed about the centre of the screen.

If the mask on the light source is removed, a clear image of the source slit should appear at the focal point of the condensing lens, and the schlieren diaphragm can be placed in position. Any masking diaphragms required near the cell can now be installed and the optical system is then set up and nominally aligned. It remains to adjust all the components to put them in their final positions.

Final adjustments

It should be remembered during the final adjustment of the optical components that they are all part of a single optical system, so that any adjustment may, to some extent, upset a previous adjustment. In the procedure outlined here, this effect has been kept to a minimum but, after adjusting the position of a component in any one direction, it is important to check that a previous adjustment in another direction has not been disturbed.

Rotational adjustment of lenses about the x and z axes

Throughout the adjustments of the various lenses it is necessary to ensure that the axis of the lens is coincident with the optical axis of the system as a whole. Lateral adjustments, except for the z-axis adjustment of the cylindrical lens, have been made during the initial setting up of the system, and if the optical axis is parallel to the optical bench, on which the components are mounted, these adjustments should not need to be made again.

The rotational adjustments about the x and z axes have also been made during the initial setting up of the system but should be checked whenever a lens is moved. This is very easily done if masks are arranged in the optical system in order to restrict the beam of light passing through the lens to the optical axis. Sufficient light is reflected from the surface of the lens to be visible on a white screen in a darkened room. When this light is reflected exactly back along the optical axis the lens is correctly orientated.

Placing the light-source slit in the focal point of the collimating lens

If a plane mirror is placed beyond the collimating lens so as to reflect the light back towards the source, an image of the source slit will be produced which will be coincident with the source if, and only if, the source is in the focal point of the collimating lens. This adjustment is made easier if an opaque bar is placed across the lens parallel to the source slit. This produces two beams of light returning towards the light source and the image is readily located as the point where the two beams converge. This adjustment may be eased still further by interrupting part of one of these beams near to the image so that, as a screen is moved through the image plane, a short image appears to move through a longer one. It will generally be necessary to tilt the plane mirror slightly so that the reflected image falls slightly to one side of the source slit, and in some designs it is possible to use the diaphragm containing the source slit as the screen on which the image is observed. The accuracy of this method is about ± 0.5 mm, which is more than adequate since the finite width of the source slit will usually cause greater deviations of the beam from parallel than will small displacements of the source from the focal point of the collimating lens.

Ensuring that the light source is on the optical axis

If the light-source slit is on the axis of the cell, then light reflected from the cell will form an image of the source coincident with the source itself. This is checked in exactly the same way as the light source was placed in the focal point of the collimating lens except that no dark bar is required over the collimating lens. It may be necessary to move the light source slightly to one side in order to find the reflected image close to the bright source itself. In analytical ultracentrifuges the reflected image is very dim, but, if the light source is supplied with alternating current, the reflected image may be caused to reveal itself by operating the rotor at a speed close to the frequency of the electricity supply to the lamp (3000 rev min^{-1} in Britain, 3600 rev min^{-1} in the U.S.A.). This causes a stroboscopic effect between the flickering of the light and the rotation of the rotor, which causes the image to flicker at a low frequency.

Focusing of the camera lens

Since a photographic plate is a two-dimensional object but the cell is three-dimensional, it is only possible to focus one plane within the cell accurately onto the plate. Since, when a ray of light passes through a cell containing a gradient of refractive index at right-angles to the direction at which the ray enters the cell its path is a parabola (Thovert 1902), it is easy to show that, if the plane halfway through the cell is focused onto the plate, then, when they reach the plate, the rays will have the same relative positions as they had when they entered the cell. Thus, if accurate registration of the x coordinates is to be ensured, the camera lens should focus the mid-plane of the cell onto the photographic plate. However, the ray that enters the cell at a point of maximum gradient of refractive index will not be the most deviated ray because it is immediately refracted away from the position of maximum gradient. A ray that enters slightly above the point of maximum gradient and is refracted through this point at the centre of the cell will, in general, be the most deviated ray. Thus the peak of the schlieren curve will appear slightly displaced from the true position of maximum gradient towards the top of the cell. This effect, which was first discussed by Wiener (1893), causes the registration of a symmetrical peak to become skew, and is called Wiener skewing. It was shown by Svensson (1954) that this effect could be eliminated by focusing the camera lens on a plane two-thirds of the way through the cell from the entrance window. Focusing on this plane, however, causes a small error in the registered positions of menisci. Svensson has also pointed out that, if the light source is not accurately at the focal point of the collimating lens, a second distortion arises, and he concluded that both these distortions disappear if a plane two-thirds of the cell thickness from the entrance window is brought to focus by the camera lens, and if the cell magnification factor is determined experimentally by placing an object in the slightly defocused middle plane of the cell.

Thus, if the primary consideration is that the optical system should register the peaks of schlieren curves in their correct positions, the camera lens should be focused on the two-thirds plane. However, when the system is used as an interferometer, there is a further reason for focusing at the mid-plane. If the light source is extended in the x direction (as it usually is in this type of interferometer), the fringes become blurred in regions where there are gradients of refractive index (Svensson 1954). This effect only disappears if the camera is focused on the mid-plane of the cell. Svensson has suggested that this blurring can be used as a very accurate method of focusing the camera lens on the mid-plane of the cell. The error in the location of the top of a schlieren curve when the lens is focused at the mid-plane is given by

$$\Delta x = \frac{a^2(\mathrm{d}n/\mathrm{d}x)}{6n}$$

where a is the thickness of the cell in the y direction (Svensson 1954). For the maximum value of gradient which most systems can register ($0 \cdot 03$ per centimetre), this error Δx amounts to about 38 μm in a cell 1 cm long. Since, in any migration experiment, it is only the change in this quantity which would be important, the effect is only likely to be significant where the very slow migration of a material with a high diffusion coefficient is being measured. For most purposes the effect is negligible, and it is preferable to focus the camera lens on the mid-plane of the cell. A more detailed discussion of these effects has been made by Ford and Ford (1964), as applied to one particular machine.

Having decided on which plane the camera should be focused, there are two simple methods of finding the correct position of the camera lens. In the first method a glass plate with fine lines scribed upon it is inserted in the cell at the required object plane. The adjustment can be made most simply by replacing the photographic plate with a ground-glass screen and observing the image of the lines on it. The ground-glass surface, of course, must coincide exactly with the position occupied by the emulsion of the photographic plate. Because of the inevitable uncertainty about this, it is better, for really accurate work, to take photographs with the lens in slightly different positions along the optical track and to pick that position which gives the best photographic image. The adjustment can be made more accurately by putting an opaque bar across the camera lens in the z direction, which causes the images of the lines in the cell to be doubled at all positions of the camera lens except the correct one. A slight variant of this procedure has been described by Gropper (1964) in which the phase plate is set at $0\cdot5°$ to the vertical. This acts like the opaque bar mentioned above except that, when the camera lens is out of focus, the images of the lines appear bent in the region of the shadow of the phase edge. The lens is adjusted until the lines appear straight. When using this first method of focusing the camera lens it is most important to fill the cell with a fluid whose refractive index is close to that which will be used in experiments with the apparatus. This allows for the foreshortening effect caused by refraction of the light at the exit windows of the cell.

The second method, which is particularly useful for the absorption system using ultraviolet light, is to make use of light reflected from a meniscus when the light source is off-axis. The light source is moved a little off-axis in the x direction towards the bottom of the cell, which is half-filled with a typical solvent. When the image of the cell is viewed on a ground-glass screen (or photographed as discussed above), if the camera lens is too near the screen, a bright line appears below the image of the meniscus. If the camera lens is too near the cell, the bright line appears above the image of the meniscus. The camera lens is adjusted until the bright line coincides with the image of the meniscus when the latter may almost disappear. The principle of this method is illustrated in Fig. 5.1, in which the two rays that just miss the meniscus are shown as continuous lines and these define the image of the meniscus on the screen. The ray that is reflected by the meniscus is shown as a dotted line, and it can be seen that this only has its correct position relative to the other two rays in the correct image plane. Since the bright line

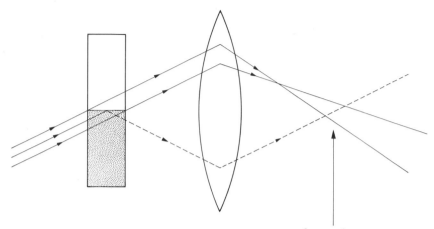

Image plane

FIG. 5.1. A very exaggerated diagram to illustrate one way of focusing the camera lens. The cell is half-filled with water (shown shaded) and three rays of light are shown approaching the cell off axis from the left. Two of these rays just miss the meniscus and continue as shown by the full lines. The third is reflected by the meniscus and follows the path shown in the broken line. It is evident that only in the correct image plane does the reflected ray fall in its correct position relative to the other rays.

only appears when the camera lens is out of focus and the light source is off-axis, the method can equally be used as a check on the position of the light source.

Positioning of the schlieren analyser

If the light source slit is rotated to the horizontal position and the analyser set to the vertical, a single dark line should appear on a ground-glass screen in the position of the photographic plate. The analyser is now rotated slowly towards the horizontal whilst observing the dark line on the screen. If the analyser is not in the correct position, the line will tilt as the analyser approaches the horizontal. In addition, the line will move in a direction at right-angles to its original position if the analyser is not in the correct position in the x direction. If the latter movement is such that the former cannot be observed, the analyser can be set to a position close to the horizontal and then adjusted in the x direction until a blurred line appears on the screen. If this is tilted, the analyser must be moved along the optical axis until the line is no longer tilted. When the analyser is in the correct position, as it is rotated from the vertical to the horizontal, the line on the screen will not move. As the horizontal position is approached, the line will appear to expand in the z direction until, when the analyser is parallel to the light-source slit, the screen will become dark. This latter condition is an extremely sensitive test of the x-axis positioning of the schlieren analyser, which position would be adequate if the line moved off the screen when the analyser was a few degrees away from the horizontal. The adjustment in the z direction is not critical. The analyser is positioned so that the line on the screen is a little above the bottom of the picture, and so that the analyser rotates about a point which is not on the optical axis.

Focusing of the cylindrical lens

The cylindrical lens must be positioned so as to give the best possible focus of the schlieren analyser. As with the focus of the camera lens, this is best done by taking photographs but may be done much more quickly and simply by replacing the photographic plate by a ground-glass screen or by using a low-powered microscope focused on the plane of the photographic plate.

A ground-glass screen is placed just in front of the schlieren analyser, which is rotated to a position that gives the sharpest image on the screen, and the cylindrical lens is then adjusted along the optical axis to obtain the clearest possible image. When the analyser is a phase plate small imperfections on the plate may be used to obtain a very accurate adjustment of the position of the cylindrical lens.

Rotational alignment of the cylindrical lens and the interference mask

For correct operation of the Rayleigh interferometer it is essential that the axis of the cylindrical lens, the light source slit, and the double-slit interference mask are all accurately aligned parallel to the x axis of the cell. The alignment of the mask over the condensing lens in the ultracentrifuge is additional to this adjustment since, in at least one design, the active slits are mounted on the cell itself and they are not adjustable. The purpose of the mask over the condensing lens is to define the position of the rotor during which the conditions for interference are satisfied. The alignment of this mask is very important since, if it is not correct, the optical system will not compare equivalent levels in the two cells. If the active slits are not placed on the cell, they may be aligned in much the same way as in a static apparatus, except that the position in the z direction is more critical because the cell will move past the slits in the arc of a circle, and, if the two slits are not symmetrically placed about a radius of the rotor, the optical system will again compare non-conjugate levels in the two cells. A thread stretched from the drive shaft of the centrifuge and passing across the centre of the lens will define a radius, and the slits can be placed symmetrically about this thread sufficiently accurately by eye. Since gradients of refractive index are often confined to one cell of the pair, the double-slit mask on the condensing lens is often offset in the z direction in order to put one slit on a radius of the rotor. This ensures that light refracted by the gradients of refractive index is not cut off by the mask. At the present point of the alignment procedure, however, the mask must be placed symmetrically about a radius. The alignment of this mask will be described first after which the general procedure can be adopted.

The principle of the method of aligning the mask and the cylindrical lens is to create exactly the same gradients of refractive index in both cells of the interferometer and to adjust the cylindrical lens until the interference fringes are straight. The requirement that the gradients of refractive index be identical in both cells is very important, and the only satisfactory way of achieving it is if a single cell that spans both interference slits can be used. In the ultracentrifuge this is easily achieved either by removing the central partition in the centrepiece of the double-sector cell that is used with the interferometer, or by placing gaskets between the centrepiece and the cell windows so that free communication exists between the two sectors, or by putting a single-sector centrepiece in the interference cell. In either case the cell is half-filled with a solution of a macromolecular solute (e.g. 0·25 per cent serum albumin solution) and the cell centrifuged at a sufficient speed for sufficient time to generate gradients of concentration at the top meniscus. With the light-source slit vertical, the cylindrical lens is adjusted rotationally about the y axis until the fringes at the photographic plate are straight at the top of the solution column. If the mask is not aligned correctly the fringes may still be curved at the bottom of the cell, but this is of no consequence. The cylindrical lens is now nominally aligned, and the rotor can be stopped and removed from the machine. The schlieren diaphragm is now aligned with respect to the axis of the cylindrical lens by placing a ground-glass screen just in front of the analyser and rotating the analyser to the position which gives the sharpest image at the photographic plate. The screen in front of the analyser may now be removed, and the mask adjusted rotationally until the fringes at the photographic plate are parallel to the image of the analyser. This adjustment is easiest if the fringes are close to the image of the analyser. With multi-slit interference sources this is almost inevitably going to be true, but with single-slit sources it may not be so. In this case the fringes must be moved closer to the image of the analyser, and this can be done either by introducing a thick block of glass just in front of the analyser and tilting it along the optical axis, or by moving any plane mirror that may be used to bend the optical system, or by moving the light source. The mask is now nominally aligned.

In an apparatus such as the Tiselius electrophoresis apparatus, it may be more difficult to place a single cell across both channels of the interferometer. However, the alignment of the cylindrical lens is likely to be less critical in this type of apparatus since, unless truly differential measurements are to be made, there is generally no gradient of refractive index in the reference channel (indeed the reference beam frequently passes through the water of the thermostat tank). In this case it will be sufficient to align the slits with the cell, and then to align the cylindrical lens with the slits by observing that position of the lens which gives the sharpest interference fringes. If, however, it is necessary to ensure that the cylindrical lens is accurately aligned with the cell it is necessary that a single cell be arranged to span both beams of the interferometer.

Having arranged a single cell spanning both beams of the interferometer, a gradient of refractive index must be created in it. In the ultracentrifuge this is easily done by filling the cell with a 0·25 per cent solution of serum albumin, or similar macromolecule, and operating the centrifuge at a low speed for sufficient time to cause some sedimentation of the macromolecule. In a diffusion apparatus a suitable boundary must be made and moved into view. In either case the schlieren system can be used to observe the gradients. If the cylindrical lens is not orientated correctly, the boundary will show as a small curve in the interference fringes, not unlike a schlieren pattern. The lens is rotated until the lines are straight when the cylindrical lens is correctly orientated.

Finally, the double-slit mask is orientated with respect to the cylindrical lens by orientating the schlieren diaphragm with the axis of the cylindrical lens, as described above, and then adjusting the mask until the interference fringes are parallel to the image of the schlieren analyser.

Rotational alignment of the light-source slit

At the end of the previous step in the alignment procedure the schlieren analyser is aligned with the axis of the cylindrical lens and its rotational scale should read 0° (or, in some designs, 90°). If the analyser is rotated through 90° and the light source set in the schlieren position, the slit and the analyser should be parallel. This is very easily checked by observing the image of the source slit on the analyser and moving the analyser in the x direction across the image. If the image and the analyser are parallel, the latter will cut the former evenly throughout its length. When the analyser is a phase plate or a fine wire and the light source is incorrectly orientated, then, when the analyser is moved through the image of the slit, a bright spot of light will be seen clearly moving along the wire. The light source is rotated until its image brightens evenly along its full length when the analyser element is moved through it. The analyser is left in the position which causes the image of the light source to fall on the edge of the schlieren analyser.

The orientation of the source slit in the interference position is best achieved by observing the quality of the interference fringes. If possible, a low-powered microscope should be set up so that the fringes may be observed directly, and the light source rotated until the best quality fringes are obtained. Alternatively, the adjustment may be made in small increments and a photograph taken after each increment. The best position is then selected from the photographs.

Positioning of the cylindrical lens in the z direction

The positioning of the cylindrical lens in the z direction is very critical if the achromatic fringe is to be observed. This can only be done with a single-slit interference source; with a multi-slit source the adjustment is less critical and the preliminary placing it probably adequate, but the quality of the fringes is the only important criterion for this purpose. With the system set up for interference, any light filter is removed, and the fringes are observed in white light. For this purpose very narrow interference slits should be used in order to obtain a wide diffraction envelope, and the fringes must either be observed through a low-powered microscope or photographed; viewing them directly on a ground-glass screen or with a simple lens of large aperture (such as is, for example, supplied with the Beckman ultracentrifuge) is seldom adequate. If the cylindrical lens is close to its correct position, one or two fringes should appear distinctly darker than the rest and the diffraction envelope should be symmetrically coloured about the centre. If the lens is sufficiently far from its correct position or too large slits are being used, the achromatic fringe may be outside the diffraction envelope. In this case, the envelope should appear orange on one side and blue on the other. The achromatic fringe is on the orange side of the envelope. Unless the envelope is very narrow so that only a very few fringes are visible within it, the absence of the achromatic fringe probably means that the rest of the system is not properly aligned. It will be worth checking the positions of the camera lens and the condensing lens. Assuming, however, that the dark fringes are visible, the cylindrical lens is adjusted in the z direction until one very dark fringe appears centred in the diffraction envelope. Strictly speaking, the achromatic fringe should be a single bright fringe, but, in the author's experience, it has proved easier in practice to adjust the system to obtain a single very dark fringe bordered by bright fringes on either side than vice versa. It also means that the achromatic fringe appears as a white fringe on a photograph, which is preferable for measurements. Finally, if the interference slits are not those which will normally be used, they should be replaced by the usual ones, and the pattern checked to ensure that the dark fringe is at the centre of the diffraction envelope.

The alignment of the system is now complete.

6

INTERPRETATION

THIS chapter will be concerned first with the measurements that must be made on the photographic records and the calculations necessary to derive the required information from them and, secondly, with the methods for calibrating the optical systems. In order to simplify the discussion it will be assumed that a single experiment has been recorded by the various modern systems. This experiment will be taken to be a simple migration experiment which has resolved two boundaries. The diagrams given to illustrate the way each system would present this result are purely schematic in order best to illustrate the principles involved. It will also be assumed that the information required from the optical record is a plot of the concentration and the gradient of concentration as a function of the x coordinate and the measurement of the x coordinate of the boundary. The way in which this information is used will, of course, depend upon the purpose of the particular experiment. Although it will be assumed that the concentrations are required in terms of weight of solute per unit volume of solution or similar units, in some common experiments it is unnecessary to convert to these units from those derived directly from the pictures.

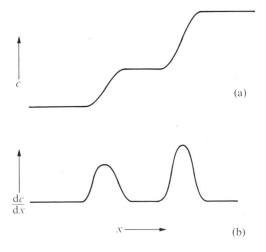

FIG. 6.1. The plots of (a) c versus x, and (b) dc/dx versus x for a typical two-boundary system. It is these curves which analysis of the optical records must yield.

   The distribution of concentration in a migration experiment after the
resolution of two boundaries is shown in Fig. 6.1(a) and the derivative of
this, which is the distribution of the gradient of concentration, is shown in
Fig. 6.1(b). It is these curves which must be derived from the optical records.
The records themselves will have an extra feature in a reference mark of some
kind from which the x coordinates are measured. In centrifuges this is usually
a hole in the rotor, and in static systems it is more commonly a wire placed
across the cell. In either case, either one edge or the centre of the image of
this hole or wire may be used as a reference. In static apparatuses it is not so
important which is used, but in the ultracentrifuge the x coordinate must
be measured from the centre of rotation of the rotor, so that the position of
the reference mark relative to the centre of the rotor must be known. In this
context it is probably better to use the centre of a hole rather than an edge
because diffraction and photographic effects can cause an edge to be slightly
indistinct. For simplicity in the discussion here the reference point will be
taken as the edge of a hole; if the centre of the hole is used it is simply a case
of measuring both edges and taking an average.

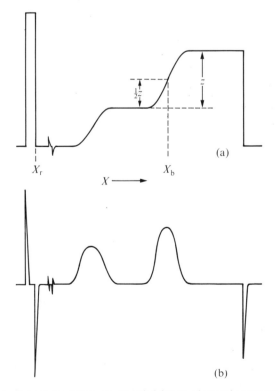

FIG. 6.2. The two-boundary system as recorded by an electronic scanner. (a) the integral
output; (b) the differential output.

Absorption systems

Electronic scanners

One of the major advantages of the electronic scanners is that they yield the required results directly. The type of trace is shown in Fig. 6.2, which is the same as Fig. 6.1 with the addition of a reference hole. Normally the scale of the z axis (extinction) will be given by the manufacturer so that extinctions can be read directly from the trace. The differential output is generally not calibrated (although there seems to be no reason for this). To determine the position of the centre of each boundary, it is only necessary to measure the distance on the chart from the centre of the boundary to the reference mark and to divide by the magnification of the system. This calculation yields the distance of the boundary from the reference mark, and, in the case of the ultracentrifuge, this is related to the position of the boundary with respect to the centre of rotation by a simple addition of the distance of the reference mark from the centre of rotation. The centre of the boundary is normally taken as the top of the peak in the differential curve (Fig. 6.2(b)). On the integral output the mid-point is usually taken as the point halfway on the z axis between the horizontal parts of the curve on either side of the boundary (see Fig. 6.2(a)). The position of the boundary is then given by

$$x = \frac{|X_b - X_r|}{m},\qquad(6.1)$$

where m is the magnification of the system.

Photographic records

The original photograph must be measured on a microdensitometer to yield a plot of the extinction of the film as a function of the x coordinate (Fig. 6.3). Since the picture is a negative the trace is inverted compared with Fig. 6.2(a) but otherwise very similar. The meniscus (if there is one) is

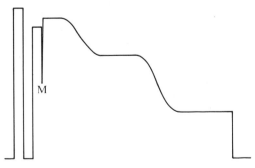

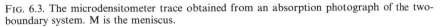

FIG. 6.3. The microdensitometer trace obtained from an absorption photograph of the two-boundary system. M is the meniscus.

indicated by a fine white line across the picture and is shown in the trace by the sharp dip at M. Assuming that the solution just below the meniscus is transparent, the blackening of the film there should be slightly greater than it is just above the meniscus. This is because slightly more light is reflected from an air–glass interface than it is from a water–glass interface. If the response of the film is linear, the curve in Fig. 6.3 is an inverted plot of the extinction of the solution as a function of the x coordinate. The distances of the centres of the boundaries from the reference mark can then be measured in exactly the same way as with the outputs of the electronic scanners described above.

In all cases the response of the film being used should be calibrated (see p. 77) under the conditions of exposure and development to be used experimentally. The response is normally as shown in Fig. 6.4; that is when the extinction of the film E is plotted against the logarithm of the product of the intensity of the light I and the time of exposure t (this product is called the exposure) the graph is curved at low exposures, becomes linear at intermediate levels, and curves again at higher exposures. Although the transition from curved region to linear region is not sharp, it is usually not difficult to choose a range over which the response is acceptably linear. So long as the whole of the trace in Fig. 6.3 (apart from the reference hole) falls within these limits the graph is indeed an inverted plot of the extinction of the solution as a function of the x coordinate. It is unlikely in practice that the upper limit will be reached as it occurs normally at extinctions of the film higher than can be measured by most densitometers. However the maximum extinction of solution which can be studied depends upon the curvature of the

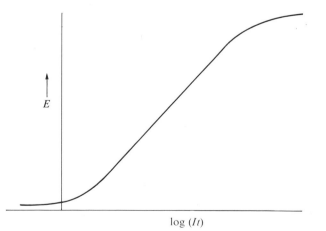

FIG. 6.4. A typical response curve for a photographic film. The ordinate is the extinction of the film as measured by a densitometer, and the abscissa is the logarithm of the product of the intensity of the light and the time of exposure. The exact form differs from one type of film to another and, for a given film, may differ for different wavelengths of light.

film response at the lower end. The position of this varies from one type of film to another.

If no acceptably linear part of the film response can be found, it is necessary to use the calibration curve of the film to correct the traces from the densitometer. This is a simple, if somewhat laborious, process. The response curve of the film is a plot of the deflection of the densitometer as a function of log (It). A level in the trace is selected as corresponding to zero extinction in the solution, and the calibration curve is used to convert the reading of the densitometer to the value of log (It) corresponding to the intensity of light incident on the solution. Call this quantity log $(It)_0$. The deflection of the densitometer at another position on the trace can now be converted to the corresponding value of log (It). Since the time of exposure is constant for all positions in a single image, the extinction of the solution is given by

$$E = \log (It)_0 - \log (It). \qquad (6.2)$$

In this way the extinction of the cell as a function of the x coordinate can be found (which will look like Fig. 6.2(a)), from which measurements can be made in the same way as from the traces obtained with the scanners. If it is only the position of the centre of the boundary which is required, this calculation need only be made for the horizontal sections on either side of the boundary, after which the calculation can be reversed to obtain the deflection of the densitometer at the halfway point in the boundary, and then the x coordinate can be obtained from the original trace. In this case, of course, one can use relative extinctions and the value of log (It) on one side of the boundary can be used as log $(It)_0$ in eqn (6.2). This procedure assumes that the illumination across the cell is uniform. This assumption is essential unless some calibrating exposure can be arranged alongside the image of the cell on the film. In this case the value of log $(It)_0$ can be deduced independently for each level.

Schlieren

The schlieren pattern obtained from a system of two boundaries in a migration experiment is shown in Fig. 6.5. The meniscus, when there is one, appears as a fine white line across the full height of the plate. The schlieren pattern itself is the white line running horizontally with the two bell-shaped curves in it. The centre of the boundary is usually taken to be at the top of the corresponding peak, and its position is measured in exactly the same way as for the differential trace obtained with absorption optics and electronic scanner. Since the picture is fairly small, the measurements are normally done with a travelling microscope, which should preferably be fitted with a projection system to throw an image of the photograph on to a screen. The magnification required is not large and such instruments are frequently called microcomparators.

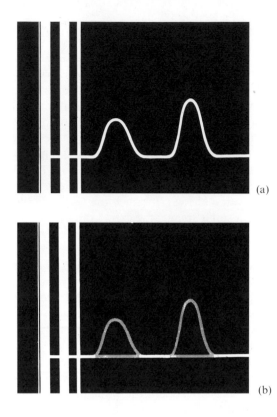

F IG . 6.5. A two-boundary system revealed by schlieren optics; (a) with a single cell and (b) with a double cell, one part of which contained a solution of uniform refractive index.

The deflection of the schlieren line away from the base line is directly proportional to the gradient of refractive index at the corresponding position in the cell. Such measurements are difficult in simple schlieren pictures, such as Fig. 6.5(a), because the position of the base line must be interpolated in the region of the peak. The problem is eased by using a double cell, similar to that used with the interferometer, but the picture then consists of two overlapping patterns, so that light from one fogs the line in the other (Fig. 6.5(b)). Consequently the contrast between line and background, which is normally so good with a phase plate, is reduced quite considerably. In Fig. 6.5(b) the base line is shown straight, but in a real case, especially in sedimentation work, it will often be curved so that the use of a double cell is almost essential.

It was shown by Thovert (1902) that the angular deviation of a beam of light passing through a gradient of refractive index is given by $a(1/n)(dn/dx)$ radians, where a is the path length through the cell and n is the refractive index of the medium. When the light emerges from the cell, it will suffer a further deviation due to the change in refractive index. If the light is incident on the exit window at angle α_1 it will emerge from that window into a medium of refractive index n' at an angle α_2 given by Snell's law as

$$\sin \alpha_2 = (n/n') \sin \alpha_1. \qquad (6.3)$$

Since in practice the angles are very small, α may be written in place of $\sin \alpha$ so that the angle of deviation of the beam becomes multiplied by n/n', and the over-all deflection of the light becomes $a(1/n')(dn/dx)$. This result is not affected by the glass window or by the water in a thermostat tank so long as all windows are plane and parallel to the walls of the cell. Since the final medium is normally air, n' is unity and could be omitted from the following equations, however it will be retained for the sake of completeness.

The displacement of the image at the schlieren diaphragm is this angle of deflection multiplied by the optical lever L, which is the distance from the condensing lens to the schlieren diaphragm. In a properly focused system, L is equal to the focal length of the condensing lens. The inclined edge of the schlieren diaphragm converts this vertical displacement to a horizontal displacement and the relationship between these two displacements is simply the tangent of the angle θ between the edge of the diaphragm and the vertical. (One manufacturer (Beckman) calibrates the angle of the schlieren diaphragm with respect to the horizontal and in this case it is the cotangent which must be used.) The deflection of the line on the photographic plate is this horizontal displacement multiplied by the magnification of the lens system in the horizontal direction, m'. Thus the over-all expression for the deflection Z of the line on the plate is

$$Z = (a/n')(dn/dx)L.m' (\tan \theta). \qquad (6.4)$$

The product $L.m'$ is a constant for a given optical system and can be determined as a calibration constant (see p. 79).

The area under the schlieren curve between two limits X_1 and X_2 is given by $\int_{X_1}^{X_2} Z \, dX$, where X is the measurement on the plate and is related to the corresponding measurement in the cell x by the magnification m of the optical system in the vertical direction. In the expression for Z in eqn (6.4), a, L, m', n', and $\tan \theta$ may be taken as constants as far as the integration is concerned so that the expression for the area A becomes

$$A = (a/n')L.m.m' (\tan \theta) \, \Delta n, \qquad (6.5)$$

where Δn is the difference in refractive index between positions x_1 and x_2, and A is therefore a measurement of the difference in concentration between two positions in the cell.

In the context of the measurements of diffusion coefficients, it is important to note that the constants relating A to Δn and Z to dn/dx differ by a factor of m. Thus, contrary to what is stated in several text books, A/Z is not directly equal to $\Delta n/(\mathrm{d}n/\mathrm{d}x)$ but to m times this quantity. However the angle θ does cancel from the expression for A/Z and so this angle can be varied during the measurement of a diffusion coefficient, without affecting the results, so long as the area is measured on every picture. It should perhaps be pointed out that the magnification factor m is referred to in the context of this book as the magnification in the vertical direction, since all directions are here referred to the cell. However, it is frequently referred to as the horizontal magnification factor since it refers to horizontal directions on the plate.

There are various ways of measuring the area under a schlieren curve. In every case the measurements must be made from the centre of the lines. This can either be judged by eye, or measurements made down each edge and an average taken. Although the latter seems a more objective way of doing it, it is much more tedious and probably no more accurate. The eye can judge the centre of a line very accurately, especially when the edges are slightly indistinct, when it judges a point half way between points of equal density on the two sides. When individual measurements are made there is a serious danger of measuring at points of different density on the two sides of the line. The photograph may be put into a photographic enlarger and an enlarged copy made from which the area can be measured directly with a planimeter, or the area may be cut out and the paper weighed, or, if the copy is made on graph paper, the squares on the paper within the area can be counted.

Alternatively the area can be obtained by numerical integration from measurements made with a microcomparator. The procedure consists of measuring Z at small intervals ΔX along the x axis. If these intervals are small compared with the extent of the integration interval, the integral is

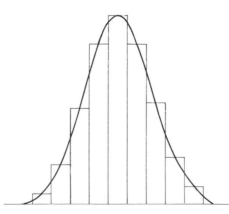

FIG. 6.6. An illustration of the approximate method of measuring the area under a peak by dividing it into a series of rectangles.

given by the product of ΔX and the sum of the values of Z obtained. For this it is necessary that the first measurement be made at $\Delta X/2$ from the beginning of the interval and the last at the same distance from the end. The procedure is illustrated in Fig. 6.6, and makes the assumption that the curve is straight in the intervals between measurements so that the small pieces outside the curve that are included are equal to the small pieces within the curve that are omitted. Obviously the smaller the interval ΔX the more nearly will this be true. A more accurate method is to use Simpson's rule. The measurements are done as above except that the first measurement of Z is made at the edge of the integration interval and there must be an odd number of measurements. If these are called $Z_0, Z_1, Z_2, Z_3, \ldots Z_n$, where n is even, the integral is given by

$$A = (\Delta X/3)\{Z_0 + Z_n + 4(Z_1 + Z_3 + \ldots + Z_{n-1}) + 2(Z_2 + Z_4 \ldots + Z_{n-2})\}.$$

$$(6.6)$$

This method assumes that the curve can be approximated to a parabola over intervals of $2\Delta X$. Although Simpson's rule is appreciably more accurate in the general case and is certainly to be recommended in experiments such as sedimentation equilibrium and the Archibald method, it is no more accurate than the simpler procedure when complete Gaussian curves are being integrated. Since most clearly-resolved boundary systems yield schlieren curves which approximate very closely to Gaussian in shape, the simpler procedure is usually adequate. For the types of curve obtained in this kind of work eleven measurements of Z and the application of Simpson's rule will yield more than adequate accuracy.

Interference

The interference picture corresponding to the two-boundary system is shown in Fig. 6.7(a), and an enlargement of one of the boundaries is shown in Fig. 6.7(b). These pictures are of the type which would be obtained with a single-slit source and interference slits a few tenths of a millimetre wide. These have been chosen rather than the clearer pictures from multi-slit sources because they introduce some complications which can be discussed, and the reader with a multi-slit source will quickly realize which parts of the discussion do not apply to him. There are two slightly different ways of measuring these pictures. Each line on the interference pattern is a graph of the concentration of the solute as a function of the x coordinate. The first method of measurement makes direct use of this fact. The second method, which in many ways is simpler, makes use of the fact that the separation of the fringes in the z direction is constant.

Having aligned the plate on the microcomparator so that the X-axis of the latter is parallel to the X-axis of the picture (fringes in reference holes at

F I G. 6.7. (a) The two-boundary system recorded by interference optics. (b) A magnification of one of the boundaries showing the points where measurements should be made.

each end of the picture are the best for this purpose, if they are available), the cross-wires are set on the centre of a white fringe to the left of the boundary and not more than halfway from the centre of the diffraction envelope towards the bottom. The reading of the Z axis of the comparator is taken, and then the plate is moved a small distance ΔX along the X axis. The cross-wires are reset onto the centre of the same fringe, by adjustment of the Z axis of the comparator, and the reading of the Z axis recorded. This process is repeated until the fringe being followed approaches the top of the diffraction envelope. Then, without moving the X axis of the comparator, the Z axis is adjusted to bring the cross-wires down onto the centre of a new fringe towards the bottom of the diffraction envelope. The new reading of the Z axis is recorded along with the number of fringes which were passed over in the process. This process is repeated every time the fringe being followed approaches the top of the diffraction envelope. The points which might be measured are indicated by crosses in Fig. 6.7(b).

When measuring pictures of this kind obtained with the ultracentrifuge, it is important not to measure fringes too close to the edges of the diffraction envelope and to keep the measurements as near the centre of the envelope as possible. This restriction, which does not apply to pictures obtained in static equipment nor so seriously to pictures obtained with multi-slit sources, arises because, in the centrifuge, the pattern has a second diffraction envelope containing no fringes superimposed upon the first. This happens because the cell is moving through the optical system and light passes through at times when the conditions for interference are not satisfied. The result is that the minima in the variations of light intensity which constitute the fringe pattern are not of zero intensity, and the over-all pattern is composed of two sinusoidal variations of light intensity. Consequently, the minima are displaced slightly from the positions they would occupy in the absence of the second diffraction envelope. The effect is fairly small and decreases as the number of fringes within the diffraction envelope increases. In a typical case, where there are twelve fringes within the envelope, the fringe halfway between the centre and edge of the envelope is out of place by about 4 per cent of the separation of the fringes. This is about the limit of the accuracy with which the fringe patterns can be read, and so, for results of the greatest accuracy, the measurements should be restricted to the central half of the diffraction envelope. In pictures obtained with multi-slit sources much more of the pattern can be used but it is best not to use the outer fifteen or so fringes where the intensity of the illumination is decreasing.

Having completed the measurements of Z at intervals along the X axis of the photograph, the difference in concentration between any two points is proportional to the difference in the Z-axis readings plus the sum of the differences between any pair of readings taken at a single X position when transferring from one fringe to another. The proportionally constant relating this to the difference in refractive index between these two levels will depend upon the particular optical system and may be determined by experiment. However, if the result of the above computation is divided by the separation of the fringes in the Z direction, a figure is obtained which is independent of the optical system and is directly related to the difference in refractive index. This is often referred to as the concentration in 'fringes' J, and is related to the change in refractive index Δn by the relation

$$J = a \, \Delta n / \lambda, \tag{6.7}$$

where a is the physical length of the cell and λ is the wavelength of the light *in vacuo*. The value of Δn can, of course, be converted to a concentration if the specific refractive increment is known (see eqn (1.2), p. 2). This method yields the concentration at any position along the cell. For sedimentation-equilibrium experiments, where the logarithm of the concentration is required for the graph from the slope of which the molecular weight is obtained, the concentrations can be left in terms of fringes, because J is directly proportional to the concentration.

The second method of measuring interferograms yields the positions on the x axis where the refractive indices differ by λ/a. The plate is aligned on the microcomparator as before and the cross-wires centred on a white fringe near the centre of the diffraction envelope to the left of the boundary. Then, leaving the Z axis fixed, the plate is moved along the X axis until the cross-wires become centred on the next white fringe below the original one as it curves upwards in the region of the boundary. The reading on the X axis is recorded, and the plate is moved on to the next fringe, and so on until the whole boundary is covered. This is illustrated by the dotted line in Fig. 6.7(b). The result is a series of X measurements at those positions in the region of the boundary where the refractive index differs by an integral multiple of

λ/a from the region to the left of the boundary. If the total change of refractive index across the boundary is required, the integral part of the result is obtained this way and the fractional part is given by AB/AC (see Fig. 6.7(b)), where A is a point at the centre of the fringe above B, and C is the point at the centre of the fringe below B, which itself is on the dotted line. Of course, this method can be used to determine the difference in concentration between any two positions in the cell, for instance the top and bottom meniscus in a sedimentation-equilibrium experiment. The result is in fringes J related to the refractive increment by eqn (6.7). When used for sedimentation-equilibrium experiments, this method tends to yield points which are concentrated more closely at the bottom of the cell than the top, whereas the first method gives more evenly spaced points. The second method also gives only as many points as there are fringes across the cell, but it is not affected by the apparent displacement of the fringes near the edge of the diffraction envelope, since all measurements are made at the same position on the Z axis. However, it is very difficult to apply a correction for any irregularities in the base line when using this second method.

The achromatic fringe

It is not possible to deduce absolute concentrations from normal interference pictures unless the concentration at one point in the cell can be deduced by other means. This information is especially necessary in the measurement of sedimentation-equilibrium experiments. The movement of the fringes, when solvent is replaced by solution in the test cell, is a direct measure of the concentration of the solution. However, all fringes look alike in monochromatic light so that, unless the solution can be introduced slowly and the movement of the fringes watched continuously, there is no way of telling which fringe on the picture taken after the solution was introduced corresponds to any particular one on the picture taken when the cell contained solvent. When using white light, however, about three adjacent fringes stand out clearly and can be distinguished in both pictures. Any one of these clear fringes may be selected for measurement so long as the same one is used on both pictures. A reference line on the Z axis is required from which the position of the fringe can be measured, and the corresponding fringe in the images of reference holes is the most suitable. The distance of the achromatic fringe from this reference line is measured on a picture taken with solvent in both halves of the cell and on a second picture taken with the solution replacing the solvent in one half of the cell. The difference between these two measurements divided by the separation of the fringes gives the concentration of the solution in fringes J related to the refractive increment by eqn (6.7). However, in practice this is not a very good procedure because, unless very narrow slits are used in the optical system, the number of lines within the diffraction envelope is rather small and limits the maximum concentration which can be measured. Moreover, it is uncertain what value of fringe separation to use in deducing J when white light is being used, since this is a quantity dependent on the wavelength of the light. The system is most commonly used in sedimentation-equilibrium experiments to determine that point in the solution column where the concentration is the same as in the solution before centrifugation.

When a uniform solution is subjected to a gravitational field in the ultra-centrifuge, initially the concentration at the top of the column decreases and the concentration near the bottom increases until equilibrium is established. If the gravitational field is not too high, there will be a point near the middle of the cell where the concentration of the solution has not changed. This is frequently called the hinge-point because, as the equilibrium is established, the interference fringes appear to hinge about this point. Consequently, at this point, the fringes are at the same level on the Z axis after equilibrium is established as they were before centrifugation. Unless the movement of the fringes can be viewed almost continuously during the establishment of equilibrium, for instance by time-lapse photography (Bethune 1970), it is necessary to use the achromatic fringe in order to identify the same fringe in the pictures taken before and after centrifugation. The technique is described by Richards and Schachman (1959) and consists of finding the point on the X axis where the achromatic fringe is at the same level on the Z axis with respect to some reference line on both pictures. Since the optical system is normally set up so that the achromatic fringe is at the centre of the diffraction envelope when there is no cell in the system, it is necessary to compensate for the difference in refractive index between solution and solvent in order to keep the fringe within the diffraction envelope. This is best done by intro-ducing a second cell identical to the first but with the solution and solvent compartments reversed. Unfortunately, this is not always possible in com-mercial instruments, in which case an inert material, such as butane-1:3-diol, must be added to the solvent in order to increase its refractive index to that of the solution. Measurements made on the achromatic fringes are usually less accurate than those made on monochromatic fringes, and it is best to use the achromatic fringe simply to find an approximate position for the hinge-point and then to repeat the measurements on monochromatic fringes. The approximate location of the hinge-point will be sufficient to identify the correct monochromatic fringe.

Calibration

In general all the optical systems require calibrating on both axes, but, as mentioned above, if movements of interference fringes are related to the separation of the fringes, the result is a quantity which is independent of magnifications in the optical system and depends only on the path length through the cell and the wavelength of light. The interference system does not therefore require calibration in the z axis. Calibration in the x direction consists of measuring the magnification factor m for this direction and the methods of doing this are the same for all the systems. This will be described first, followed by the calibration of the z axis for scanners, photographic absorption systems, and the schlieren system.

x axis

Although it is possible, at least in theory, to calculate the magnification of an optical system from the positions of the lenses and the lens formula, it is preferable to determine it experimentally. This is done by placing in the position of the cell a transparent plate on which is scribed two fine lines parallel to one another, and a centimetre or two apart (depending on the size of the cell), and a third line exactly at right-angles to them. The third line is used to align the plate accurately with the x axis of the cell, and it is then photographed through the optical system. With electronic scanners it will be necessary to supply a suitable reference cell and a scan will yield a trace with two sharp spikes on it corresponding to the two parallel lines in the cell. In commercial versions of the ultracentrifuge a balance cell is provided with two holes for reference purposes, and these may be used in place of the scribed plate described above. The distance between the parallel lines is measured in the x direction both on the photograph and on the plate itself, using the third line as a guide for direction. Division of the former measurement by the latter gives the magnification factor directly. In the case of photographic absorption systems the photograph should be scanned by the microdensitometer and measurements made from the trace so obtained in order to take into account the magnification of the densitometer.

Scanners

Commercial versions of the electronic scanners should not need independent calibration of the z axis, since this can be designed into them and should be given by the manufacturer. However, for technical reasons in the ultracentrifuge, it is usually necessary to use light with a fairly large bandwidth. This results in a failure of the Beer–Lambert law which appears as curvature of the response of the scanner. The only really satisfactory way to callibrate the scanner is to put solutions of known extinction into the machine and to use the scanner to measure their extinctions and to plot the deflection of the pen on the recorder as a function of the known extinctions of the solutions. If measurements are required at high extinctions (say greater than unity), it becomes important to check the linearity of the response because stray light can easily cause serious errors at the top end of the response. It is probably better to use a shorter cell and so reduce the extinctions to be measured.

Less laborious methods are available if the necessary electronic test equipment is at hand. The detector is disconnected and replaced by a pulse generator which generates pulses of known amplitude and duration which mimic those that would be obtained from the detector under normal operating conditions. It is necessary to know the relationship between the output of the detector and the intensity of the light falling on it. Although photomultipliers have a linear response to light this is not necessary for linear operation of the scanning system as a whole. Tests of this sort check only the operation of the scanning system itself, and they do not replace the test with solutions in the cell, because only the latter will reveal any nonlinearity caused by stray light. However, once the response line is determined electronically, the effect of any stray light can be checked with a single solution whose extinction is as high as it is required to use. If the deflection of the recorder for this solution falls on the calibration line the stray light may be taken as unimportant.

There is a simple way of checking the linearity of the over-all system and also the detector alone. If the cell is filled with a solution of extinction about 0·3 this will transmit about 50 per cent of the light transmitted by the reference cell. The recorder will indicate the actual extinction. If the intensity of the light incident on the cells is changed the deflection of the recorder should remain constant. The systems using a photomultiplier may have a feedback circuit which alters the high voltage on the photomultiplier in such a way that the size of the reference pulse from it remains constant when the incident light level changes. It would be necessary to disable this circuit, but with these systems it is possible to do the test by leaving the light intensity constant and altering the voltage applied to the photomultiplier. The reading of the recorder will only remain constant in this test if the over-all response of the system is linear. To test the linearity of the detector, the pulses at the output of the linear stages of amplification are measured on an oscilloscope. The intensity of the light is then reduced and the decrease in the size of the pulses is compensated by adjusting a gain control on the amplifier. If both pulses return to their original size together, the response of the detector and the amplifiers together must be linear. This method will not detect nonlinearity due to stray light.

Photographic absorption systems

Calibration of absorption systems using photographic detectors consists of determining the response curve of the film being used at the wavelength being used. This is a plot of the deflection D of the pen on the microdensitometer used for scanning the film as a function of the logarithm of the product of the intensity of the light I and the time of exposure t, and is of the form shown in Fig. 6.4. Since the abscissa $\log(It)$ may be written as $\log I + \log t$, the same curve will be obtained by plotting D against either

$\log t$ at constant I, or $\log I$ at constant t. (It should perhaps be admitted that this statement is not strictly true owing to an effect known as reciprocity failure, but the effect, small in any case for the range of exposures used in this sort of work, can be made negligible by suitable choice of photographic material.) With the cell full of a transparent solution, a series of photographs is taken using exposures from just above the length of time which will be used in the experiments down to about one twentieth of this. The film is then developed, fixed, washed, and dried in the standard way, and the blackening is measured on the densitometer. The deflection of the pen D is plotted against the logarithm of the time of exposure. This is the calibration curve for the film. The abscissa, which was plotted as $\log t$, can be used directly as a scale of $\log I$ because, for measurement of extinction, the units in which I is measured are not important and the response of the film depends on the product of I and t; this product varies in the same way whichever parameter is varied, if the other is kept constant. The actual intensity of light giving rise to a deflection D on the densitometer is given by $I_0 t/t_0$, where I_0 is the intensity of the light incident on the film during the calibration exposures, t is the time of exposure during the calibration exposures that would have given a deflection of D on the densitometer, and t_0 is the actual time of exposure given to the final photograph. It is obvious that when the ratio of intensities is taken for calculation of an extinction the factor t/t_0 cancels out.

If, under a particular set of conditions, reciprocity failure is thought to be significant, there is no alternative to calibrating the film by putting solutions of known extinction in the cell and taking a series of exposures. In this case, it is imperative to have a record of the intensity of the incident light with each picture, because the process takes quite a long time and variations in the intensity of the light source between exposures would invalidate the results. An alternative is to place in the cell a plate having a series of areas of different transmission and to photograph this onto the film. The plate can be a photographic plate blackened by a series of exposures or a glass plate onto which layers of aluminium have been evaporated. This has the advantage that the calibration is done in a single exposure, but the transmission of the calibration plate must be measured independently. Some care is needed with this because the absorption optical system is sensitive to position, whereas an ordinary spectrophotometer, in which the calibration measurements might be made, is not. Unless the area is very evenly blackened this can cause some error. The author has also found some significant discrepancies between the apparent extinction of pieces of blackened film measured in a standard spectrophotometer, on the one hand, and the absorption system of an ultracentrifuge, on the other. These discrepancies remain unexplained but may be caused by the different geometry of the two optical systems and the fact that a part of the apparent absorption was caused by scattering of the light from the film rather than true absorption, and one optical system probably collected more of this scattered light than the other.

Schlieren

Calibration of the z axis of the schlieren system consists of determining the product $L.m'$ in eqn (6.4). The second calibration constant, which relates areas to differences in concentration, is $m.L.m'$, where m is the magnification factor in the x direction, which may be determined as described above, and L is the distance from the principal plane of the condensing lens to the schlieren diaphragm. This distance can be measured directly if the principal plane of the lens can be determined from its geometry (it will normally be close to the centre of the lens), although there may be some practical difficulty if the optical system is bent, which is often the case. The magnification factor in the z direction (m') can be determined by replacing the schlieren diaphragm by a glass plate similar to that described above for determining the magnification factor in the x direction. The plate is rotated to obtain the sharpest image of the lines at the photographic plate (that is to ensure that the lines are parallel to the axis of the cylindrical lens) and a photograph taken. When doing this it will be necessary to defocus the light falling onto the plane of the schlieren diaphragm, which is most easily done by placing a ground-glass screen just in front of that plane. The magnification factor can then be determined directly from measurements of the separation of the lines on the plate and on the image. This might be called the mechanical method of determining the calibration constant.

In many ways it is probably preferable to use an experimental method based on deviations of light in an experimental cell, because this will not involve any interference with components in the optical system after it has been aligned (see Chapter 5). One manufacturer (Beckman) can supply a special calibration cell for the calibration of the optical system in their ultracentrifuge. This consists of a cell which fits into the hole in the rotor and which contains a quartz prism aligned with the x direction and two fine lines at right-angles to this. The distance between the lines and the angular deviation of the light caused by the prism are given by the manufacturer. When this cell is photographed through the optical system, it produces a line displaced upwards on the photographic plate. This deflection Z is measured at some convenient angle θ' of the schlieren diaphragm with respect to the horizontal (Beckman calibrate the angle of the schlieren diaphragm with respect to the horizontal). The calibration constant is then given by

$$L.m' = (Z \tan \theta')/\phi, \qquad (6.8)$$

where ϕ is the angular deviation in radians of the beam of light caused by the quartz prism.

An alternative method is to create a boundary in the cell between a solution of known concentration and its solvent. The boundary is allowed to diffuse until a convenient bell-shaped curve is obtained on the photographic plate. The area under this curve is measured from which the calibration constant $m \cdot L \cdot m'$ can be obtained from eqn (6.5) and m can be determined as described above. The quantity Δn in eqn (6.5) must be evaluated separately either in a differential refractometer or from the concentration for the solute. Solutions of potassium chloride or of sucrose are suitable for this purpose, and the refractive indices of such solutions are available in the literature (e.g. Kruis 1936; Browne and Zerban 1941; Gosting and Morris 1949).

7

COMPARISON OF THE SYSTEMS

ALTHOUGH the advantages and disadvantages of the three more important optical systems have been listed at the end of their respective chapters, their relative merits will be discussed here in more detail under various objective headings in the hope that this will help the reader to select the most suitable system for his own particular needs. The optical systems are designed to measure three things, concentrations of solute in the experimental cell, gradients of this concentration with respect to the x coordinate, and the x coordinate itself. The aspects of these systems which must be considered are the accuracy with which they can measure these parameters, their sensitivity both to small absolute values of them and also to small changes in larger values, the maximum values they can record, their ability to discriminate between different solutes, their convenience in use, and the cost of installing the system in the first place and of using it later. Much of the discussion in this chapter is based on the author's personal experience which does not cover all the commercial systems available, and the conclusions must be influenced by the quality of the lenses used and the precision with which the whole optical system has been designed and laid out. These considerations are likely to differ between the different manufacturers and are not generally under the control of the individual experimenter unless he is in the fortunate position of being able to choose a new instrument, in which case considerations other than the quality of the optics are also going to influence his choice.

Range

This section is concerned with the maximum values of the variables that the optical systems can record. The smallest values are discussed under the next heading.

Concentration

The upper limit of concentration that can be registered by the absorption system depends upon its ability to respond to the very low levels of light transmitted by highly absorbing solutions. However, beyond an extinction of unity, purity of the light source becomes an important consideration. As the intensity of the transmitted light falls so the contribution of stray light, which may not be absorbed by the solution, becomes more and more significant. The result of this is that the response of the system becomes non-linear. With photographic systems this is likely to be less important than with the photoelectric scanners, because the photographic response is in any case frequently not linear in this region and calibration will be necessary. This calibration will allow for any stray light so long as the calibration is done with the same system and at the same nominal wavelength as the final experiment. Photoelectric scanners can be made linear up to extinctions of about 2, and, at this level, stray light is likely to be the limiting feature. The levels of stray light will depend not only on the individual design but also on such things as the cleanliness of the optical components which change with time. Monochromators with their entrance and exit slits close to one another are particularly prone to this trouble, because the common colli-mating mirror (if there is one) becomes cloudy with time and reflects light directly from the entrance slit to the exit slit. This sort of trouble is likely to differ in magnitude at different wavelengths, so that if the stray light is satis-factory in the visible range it may not be so in the ultraviolet.

The refractometric optical systems measure differences in concentration between different parts of the cell. Consequently the absolute values are not important (unlike the absorption system), but there is, of course, a limit to the size of concentration change which the system can measure. In both systems the limitation is the size of gradient of refractive index which they can handle and the distance over which this can be maintained. There is a further restriction on the interferometers. As the concentration difference between the solution and the reference solvent increases, the visibility of the fringes decreases because the light sources in general use at present are not truly monochromatic but consist of a narrow band of wavelengths. Thus the fringes observed are not a single set of fringes but are the superposition of a large number of sets with very slightly different spacings. As one moves further from the point of equal path length for the two beams of light, these sets of fringes become more and more out of phase with one another, so that the contrast between the bright and the dark fringes decreases and the fringes eventually disappear. An inert material can usually be added to the reference solvent in order to increase its refractive index to bring it closer to that of the solution under study, as is done when the achromatic fringe is being used. The use of a laser for the light source would improve the visibility of the fringes under these conditions (Paul and Yphantis 1972).

Gradients of concentration

The absorption and interferometric systems measure concentrations and, therefore, they yield information on gradients of concentration only after differentiation. This normally means drawing tangents to experimental curves or using finite differences over small differences in x. The electronic scanners normally have electronic differentiating circuits which perform this function automatically, and these will have an upper limit depending on the detailed design of the circuits. The interferometer has an upper limit set by the fact that, as the gradient becomes greater, the fringes become very closely spaced in the x direction and may eventually become so close that the photographic plate is no longer able to resolve them.

The schlieren optical system registers gradients of refractive index directly, but the maximum value it can register depends upon the construction of the particular optical system and no general figure is possible. The limit is caused by the fact that as the gradient of refractive index becomes larger so also does the deflection of the light as it passes through the cell. Eventually this deflection becomes so large that the deflected light misses one optical component (usually the camera lens) and so never reaches the photographic plate. This is registered as a black band across the picture covering all regions where the gradient of refractive index exceeds the limit, and this corresponds exactly to the old Toepler system (Tiselius *et al.* 1937, see p. 23). The same limitation applies to all optical systems and, in connection with absorption optics, it is as well to realize that, for a given gradient of concentration, gradients of refractive index are greater for ultraviolet light than for visible light. If regular work with solutions giving high gradients of refractive index is being undertaken, compensation for the deflection of the light is possible either by using prismatic windows on the cell or by setting the optical components off-axis, and this could double the maximum gradient which can be registered. The commercial schlieren systems are able to register gradients of refractive index up to about 0.03 per centimeter in a cell 1 cm long.

Sensitivity

The sensitivity of an optical system can be gauged either in terms of the minimum value of concentration or gradient of concentration which it can detect or in terms of the smallest change in these variables which it can detect in larger over-all values. In most systems these two are comparable and need not be distinguished. However, with the absorption system using electronic detectors, the sensitivity of the system must normally be set up so that the maximum extinction being measured is on the scale of the final recorder, and, if this extinction is large, the recorder may be unable to register a very small change. It is sometimes possible to offset the recorder so that the base line is off the scale, and then large extinctions can be measured on high sensitivity settings.

Concentration

Since the schlieren system records gradients of concentration rather than concentrations themselves, the minimum change in concentration which it will detect is very dependent on the extension of the change in the x

direction. Thus a very small step in concentration is detectable as a fine line across the picture, because a step represents an infinitely high gradient of refractive index. On the other hand a change of as much as 1 mg ml^{-1} spread out over a centimetre in the x direction would probably not be detected. This same condition applies to a lesser extent to the absorption and inter-ference systems because, the lower the rate of change of concentration, the more important do small variations in the base line become. In general the interference system is incapable of measuring a sudden change in concen-tration, because it is not possible to follow a single fringe through the dis-continuity unless the achromatic fringe is being used. Otherwise, a change in concentration corresponding to a change in optical path length in the cell of one wavelength of light is readily detectable, and this may be taken as the sensitivity of the system. For a cell 1 cm long and green light, this corresponds to a change in refractive index of about 5×10^{-5} which, for a typical protein, corresponds to about 0.3 mg ml^{-1}.

It is more difficult to estimate the smallest change in concentration which could be detected by the absorption system since so much depends on the detector in use and the care exercised by the experimenter. Under ideal conditions the photographic system could probably detect a change in extinction of 0.02 but it is unlikely that this could be realized on a routine basis, and such a small change would frequently be lost in background irregu-larities caused by the single-beam nature of the system. Electronic scanners are much better in this respect, although it is doubtful if they could detect a change of less than 0.01 reliably. In routine use changes of extinction of 0.1 and 0.05 probably represent a reasonable estimate of the sensitivity of the of the photographic and photoelectric systems respectively.

Gradients of concentration

Since the absorption and interferometric systems are designed to measure concentrations rather than gradients of concentration, it is almost impossible to estimate how small a gradient might be detected by these means. So much will depend upon the accuracy with which the base line can be determined. The commercial versions of the electronic scanners could probably only detect a gradient of about 2 units per centimetre on their differential outputs. These are probably designed primarily to facilitate the location of boundaries in migration experiments rather than for detection of small concentration gradients.

The sensitivity of the schlieren system can to some extent be adjusted by rotating the schlieren analyser. Although, in theory, the sensitivity can be adjusted from zero to infinity by adjusting the angle of the analyser between $0°$ and $90°$ to the vertical (x direction), in practice there is little point in setting the angle below about $20°$, when the most deflected rays that the system will accept are registered on the plate. Although the magnification in the

z direction increases as this angle is increased, the thickness of the line increases also, and, if the minimum detectable deflection is said to be equal to the thickness of the line, this minimum deflection does not alter very much as the analyser is rotated. With modern phase plates this would correspond to a minimum gradient of about 3×10^{-3} cm^{-1}.

x coordinate

The sensitivity of a system to the x coordinate would normally be called the resolution of the system. Purely optical limitations in this respect are limited to diffraction effects and imperfections in the optical components. These will normally be within the resolution of the photographic plate so this may be taken as the limiting factor. However, imperfect alignment of the system, in particular the alignment of the light source on a line at right angles to the gravitational field at the cell, will degrade the resolution in the x direction. It must also be remembered that the cell is a three-dimensional object, whereas the photographic image is two-dimensional. Consequently only one plane in the cell can be faithfully registered. This point is discussed in more detail under the heading *Accuracy* (p. 91).

A further limit to resolution is introduced by the electronic scanners used with the absorption system. The cause of this is fairly obvious with the television system, but applies equally to the photomultiplier systems. In the television camera the continuous image is scanned in a series of horizontal lines, and it is obvious that any detail smaller than the distance between these lines cannot be resolved. In the system described by Lloyd and Esnouf (1974) each frame consists of 625 lines, but fifty of these are blanked to allow for field flyback so that only 575 lines are used to record the picture. In television terms this would be referred to as 575-line resolution. In the ultracentrifuge the height of the cell is about 16 mm so that this resolution corresponds to about 28 μm in the cell. This resolution is independent of light intensity and scanning speed. The equivalent limitation in the photomultiplier systems lies in the narrow masking slit over the photomultiplier. The narrowest slit mentioned by Lamers *et al.* (1963) was 37 μm wide and, since they used a magnification between cell and image of about 2, this corresponds to a resolution in the cell of about 18 μm. However, they also state that they generally used a slit 75 μm wide corresponding to a resolution of about 35 μm. In this design the size of slit that is necessary depends on the intensity of the illumination available, and hence the resolution is likely to vary with the wavelength and bandwidth of the incident illumination. It is evident that the two systems have about the same resolution. Spragg *et al.* (1965) do not give the size of slit they used in their design. It should perhaps be mentioned that in order to achieve this resolution the optical system must be very accurately focused. This point is discussed in more detail on p. 91.

Accuracy

Concentration

Only the absorption system enables absolute values of concentration (in terms of extinction) to be measured directly. The interferometer can be used for this purpose if the achromatic fringe is used, but it is generally used to measure differences in concentration between two points in the cell. The schlieren system can measure only such differences. The accuracy of the photographic absorption system depends largely on the reproducibility of the exposure (which depends upon the intensity of the light source) and the developing conditions. By using stable light sources, carefully defined exposure times, and carefully controlled developing conditions (time, temperature, and state of the developer), the same part of the over-all film response curve can be used each time, and the exact shape of this curve can be determined for the particular film being used. Alternatively, it is possible to arrange a calibration exposure together with each experimental exposure so as to have an internal calibration of the film response. A special hole for use in an ultracentrifuge for this purpose has been described (Robkin, Meselson, and Vinograd 1959), and in static apparatus it is easy to arrange for a calibrating wedge to be photographed along with the cell. If this wedge is arranged to lie alongside the cell, unevenness of the illumination can largely be allowed for. A major difficulty of this technique is that the system uses a single beam of light so that irregular absorptions within the system are recorded by it and are difficult to eliminate. One of the major advantages of the electronic scanners is their ability to subtract automatically the absorption of a reference cell so that these irregular absorptions and any unevenness of illumination are compensated. Neither method is as accurate as a good static spectrophotometer but, with care, either can be as good as a simple recording-type spectrophotometer. In the author's opinion scanners are as yet by no means fully developed, and interesting developments include connecting them on line to a digital computer (Spragg and Goodman 1969). An accuracy of ± 0.01 in extinction is probably a reasonable estimate of the present state of these systems, but it should perhaps be mentioned that one manufacturer (M.S.E.) offers full-scale deflection on the recorder (about 9 cm) for an extinction of 0.25, which suggests that the chart could be read to smaller intervals than suggested above. The accuracy of the photographic system is not as high as that of the scanners mainly because of the single-beam nature of this method.

One further word of warning is necessary when estimating the accuracy of any absorption system for absolute concentration determinations. It is most important that the trace contains a region of zero extinction so that the position of the base line is indicated. In the ultracentrifuge there will generally be an air-space above the solution which can be used for this purpose but this might not be true for other systems. The base line produced by a scanner in the absence of any pulses into the measuring system is not necessarily the same as the level corresponding to zero extinction of the solution under test if, for some reason, the two cells are not exactly matched for background absorption.

The Rayleigh interferometer produces curves of concentration versus the x coordinate directly. These patterns can be measured to an accuracy of about ± 0.02 of a fringe spacing and a movement of one fringe spacing corresponds, in a cell 1 cm long, to a change in refractive index of about 5×10^{-5}. Thus changes in refractive index can be measured to about $\pm 10^{-6}$. If the deflection of the achromatic fringe was being used to measure concentration directly, most optical systems could show a movement of only about four fringe spacings, so that an accuracy of 0.5 per cent is the maximum possible accuracy. By using very narrow slits in the interferometer it is possible to increase the number of fringes within the diffraction envelope with a proportional increase in the maximum attainable accuracy. It should be pointed out, however, that the measurement also requires knowledge of the path length within the cell and this will seldom be known with an accuracy much better than 0.5 per cent. The best way of measuring concentrations with the interferometer is to make use of its ability to measure differences in concentration and to arrange for a region within the cell where the concentration is zero. This can be done easily by forming a boundary between the solution and the reference solvent. This produces a picture similar to Fig. 4.8, and a count of the fringes across the boundary yields the concentration of the solution. This count could be made to an accuracy of ± 0.02 of a fringe, and there could be a large number of fringes across the boundary, so long as the gradient of refractive index is not too high and the cell is long enough to accommodate the entire boundary. There are two limits to this procedure. The first is the requirement to know the path length through the cell, as mentioned above. The second is that the visibility of the fringes becomes poorer as the path length between the two rays that form them increases (see p. 82). The use of a laser for the light source (available from one manufacturer) would improve this situation.

The above discussion implies that the interferometer is capable of measuring concentrations to a very high order of accuracy indeed. This is true, within the limitations mentioned above, for measurements between two levels where there is little or no gradient of the refractive index. However, in regions where there is such a gradient, application of eqn (6.7) would result in error. It has been shown by Svensson (1954) that the total difference in optical path length ΔS for two rays passing through media of refractive index n and n_0 in the Rayleigh interferometer is given by

$$\Delta S = a(n-n_0) + \frac{a^2 \beta (\mathrm{d}n/\mathrm{d}x)(1-2r)}{2n_0} + \frac{a^3 (\mathrm{d}n/\mathrm{d}x)^2 (2-3r)}{6n_0}, \qquad (7.1)$$

in which n_0 is the refractive index of the reference solution, β is the angle at which a ray of light is incident on the cell (in the xy plane and with respect to the optical axis), and the camera lens is focused on a plane at a distance ra from the entrance window of the cell. The last two terms in eqn (7.1) represent deviations from eqn (6.7). The first of these disappears if either (dn/dx) or β is zero or if the camera lens is focused on the mid-plane of the cell ($r = 0.5$). β can only be zero for all rays entering the solution if the light source is not extended in the x direction, which means it would have to be a point light source, and this is why the fringes become blurred if an extended light source is used and the camera lens is not focused on the mid-plane of the cell. In a normally focused system, therefore, it is the last term of eqn (7.1) which represents the deviation from eqn (6.7). In a cell 1 cm long, this term has a value of 0.02 λ when the gradient of refractive index is 4×10^{-3}, which corresponds to 76 fringes per centimetre at the cell (or $76/m$ fringes per centimetre on the photographic plate). Thus if the concentration is required at a point where the gradient of concentration exceeds this figure, which is well within the limits of resolution of a photographic plate, the error incurred by applying eqn (6.7) will be detectable.

It is not possible to measure absolute concentrations using schlieren optics but it is possible to measure differences in concentration. Absolute values can, of course, be obtained by the boundary technique described above. The difference in concentration between any two levels in the cell is proportional to the area under the schlieren curve between these two levels. It helps this measurement considerably if the base line is included in the pattern, and this can usually be arranged, albeit at some loss of contrast in the picture, by arranging a reference solution in the light beam alongside the test solution, as is done with the interferometer. Another way of obtaining the base line is to have an extra wire across the schlieren analyser at right-angle to the image of the light source, but this should be built into the analyser for best results and usually it is not. Although the schlieren line can be quite narrow when using a phase plate and a heavy exposure of the photograph, this is not possible when using the double cell for obtaining the base line, and generally the thickness of the line and its indistinct edges limit the accuracy. Although very careful measurement can give the area under the curve accurate to a few per cent, this area is related to the concentration by a calibration constant which itself is best obtained experimentally, and the method cannot be recommended for accurate measurements of concentrations.

It should, however, be noted that there is a further aspect of the schlieren system which was pointed out by Antweiler (1951). The schlieren system from the light source to the schlieren analyser is essentially the same as the Gouy interferometer (see p. 50), so the band of light produced in the plane of the analyser is broken up into a series of interference fringes. When these are projected through the astigmatic camera, the schlieren line appears broken up into a series of fine dots. Each of these represents a change of refractive index in the cell of λ/a, where λ is the wavelength of light *in vacuo* and a is the physical path length of the cell. Antweiler claims that by counting these dots concentration changes can be measured to an accuracy of λ/a, whereas Svensson (1939) claims a limit of $9\lambda/a$ for simple area measurements; this may be compared with $0.02\ \lambda/a$ for the interferometer. Both these authors used a narrow slit as the schlieren analyser and the effect only appears with analysers of this kind.

To summarize this discussion, the most accurate means of measuring concentrations is by interferometry, followed by absorption optics with an electronic scanner or, with more trouble, photography. Other considerations may, of course, dictate the system to be used and, with care, even the schlieren system is capable of an accuracy sufficient for many purposes.

Gradients of concentration

Only the schlieren system measures gradients of refractive index directly. When using a modern phase plate it is possible to measure large deflections of the schlieren line to better than 1 per cent. For smaller deflections the apparent accuracy decreases, because the uncertainty of the position of the line remains constant, independent of the deflection. However, because the schlieren optical system, from the light source to the schlieren analyser, is identical with that of the Gouy interferometer, the band of light at the analyser is broken up into a series of interference fringes. It has been shown by Kegeles and Gosting (1947) that this effect causes the deflection of the schlieren line to be slightly smaller than it should be, and that this error increases with the deflection. The error, which amounts to about 2 per cent, causes the peaks in schlieren optics to appear slightly flattened. In order to convert the deflection, as measured, to the value of the gradient of refractive index, it is necessary to use eqn (6.4), and it is unlikely that the other quantities in this equation will be known to an accuracy better than about 1 per cent. The over-all uncertainty in the derived value of dn/dx is therefore likely to be between 2 per cent and 3 per cent.

Although the accuracy of the Rayleigh interferometer is much greater than the schlieren system, the difficulties of differentiating the integral curves are such that it is probably no more accurate than the schlieren system for the measurement of gradients. The modifications to the interferometer for obtaining derivative curves directly (Svensson 1950b; Wiedemann 1952) are very much more accurate, and Svensson claims an accuracy of 0.2 per cent under ideal conditions.

The differential outputs of the electronic scanners used in conjunction with the absorption optical system also yield direct measurement of the gradient of the concentration. The accuracy of analogue electronic circuits is such that this system could be at least as accurate as the schlieren system. In the television design of Lloyd and Esnouf (1974) the differentiation is done by taking the finite difference between successive lines of television scan. Within this limitation the differential output is as accurate as the integral output.

x coordinate

The accuracy with which a system can register the x coordinate depends on the quality of the optical components to such a degree that no general figure is possible. It will normally be safe to assume that the measurement of the x coordinate is more accurate than any other measurements with the optical system. There are, however, a few possible exceptions. The error in the location of the centre of a boundary due to Wiener skewing of the pattern has already been discussed in Chapter 5 (see p. 56) and applies to all three optical systems. A possible further source of error can arise with the electronic scanners, because the final graph is plotted against time and only relates linearly to the x coordinate if the movement of the photomultiplier (Lamers et al. 1963), the movement of the image (Spragg et al. 1965), or the movement of the electron beam in the television camera (Lloyd and Esnouf 1974) is at an absolutely constant speed. In the mechanical designs this can be ensured by using synchronous motors whose speed is locked to the frequency of the electricity supply, and in the television camera the scanning circuits can be made to give a linear scan to within a fraction of 1 per cent.

Another source of error is present in systems used with an ultracentrifuge. In this application the x coordinate has to be measured from the centre of rotation of the rotor. This necessitates a reference mark of some kind mounted in the rotor, and whose distance from the centre of the rotor is known. In addition, modern centrifuges use self-balancing rotors, which spin about their own centre of gravity; if the rotor is not perfectly balanced (and this type of rotor is used so that exact balancing of the rotor is not needed for safe operation), the centre of gravity may not coincide with the geometrical centre of the rotor and may vary very slightly from one experiment to another. This effect is likely to be very small but is present nevertheless. A more serious source of error arises because the cell moves through the optical system in the arc of a circle, and only the absorption system and the Jamin interferometer are able to register this fact. In the schlieren and Rayleigh systems, the cylindrical lens can only be aligned correctly with the cell when the latter is in one particular position. Masks are used over the collimating and condensing lenses to restrict the part of the circle in which the cell is visible, but these cannot be made infinitely narrow. This effect is most serious in the Rayleigh interferometer because there are two cells mounted in the rotor and these cannot be parallel to one another. The mask that is mounted on the collimating or the condensing lens must have two parallel slits, and both of these cannot be on a radius of the rotor. If they are arranged symmetrically around the radius, the meniscus and the reference mark are clearly seen, but they are not in their correct positions because the cells are not aligned correctly with the axis of the cylindrical lens at the moment when interference conditions are satisfied. The effect is to change both the x coordinate of the reference mark and the apparent magnification factor. If the off-set arrangement is used so that one slit is exactly on the radius of the rotor, a meniscus or reference mark has a slightly different x coordinate as it passes the offset slit to that which it had when it passed the radial slit. In an accurately focused system this effect is clearly visible if the offset double-slit mask is used with schlieren optics; two images of the meniscus are clearly visible. Although the solution is not to use the double-slit mask when using schlieren optics, with some machines this is not possible if both sets of records are needed in the same experiment. Also, the effect is still present when using the interferometer, which resolves the meniscus and reference mark rather badly for this reason.

As already mentioned above, the accuracy of the registration of the x coordinate will be affected by the focusing of the camera lens and the positioning of the light source. The former point is discussed fully in Chapter 5, but it is obvious that, if the camera lens is not focused accurately, the resolution of the x coordinate will be degraded. There will also be a decrease in the resolution if the light source is not on a line at right-angles to the gravitational field at the cell. This arises because the solution consists of a series of thin layers lying one on top of the next, and these cannot be resolved unless the light passes through exactly parallel to the layers. It is evident that any effect which degrades the resolution of the system will affect the accuracy with which the x coordinate can be measured.

Discrimination between solutes

Since virtually all solutes dissolved in a solvent cause a change in refractive index, and since the magnitude of this effect and the variation of this magnitude with wavelength are very similar, the refractometric systems are intrinsically unable to distinguish one solute from another or to detect small quantities of one solute in the presence of large quantities of another. It is, of course, possible in a migration experiment to detect a component, present in small amount, in the presence of a larger amount of another component, if the boundaries of the two components become well separated. However, it is not possible to follow the migration of the two independently of one another; the pattern must always show both components.

The ability of the absorption system to distinguish one solute from another is perhaps its greatest single advantage over the refractometric ones. The two components must, of course, have different absorption spectra so that wavelengths can be found at which their relative absorptions differ. Unfortunately, most proteins, for example, have very similar absorption spectra, because the absorbing groups are the same and one spectrum may be considered as a magnified version of another. This means that, in general, even the absorption system cannot examine one protein in the presence of another unless one is abnormal in having an absorption band outside the normal range of 200–300 nm (e.g. haemoglobin) or in having an abnormally low absorption in the 280 nm band (e.g. protamine or some histones). It is possible, if the proteins are available purified, to label one with a group absorbing at a wavelength above 300 nm. Since the absorption of proteins in the region of 200–220 nm is caused by the peptide bond itself, all proteins have very similar extinction coefficients in this part of the spectrum. At 280 nm, however, the absorption is caused by the aromatic amino acids contained in the protein, and the proportions of these vary, so that the ratio of the extinction at 210 nm to that at 280 nm is different for different proteins. Unfortunately, no absorption system at present available will work at wavelengths as low as 210 nm, and, in any case, the ratio of the extinctions at these wavelengths would be very large and consequently difficult to measure.

Convenience

Undoubtedly the most convenient system to use for routine measurements is the schlieren system. It produces patterns which are directly visible on a ground-glass screen or through a simple viewing lens and which are very readily interpreted by the eye in terms of the number of components present. In many cases it may be possible to obtain the required information directly by eye without the need to take photographs at all. The differential form, producing as it does discrete peaks for each boundary present, seems to be particularly pleasing to the eye and makes the recognition of overlapping boundaries easier. However the resolution of two discrete peaks sometimes implies to the uninitiated that the components of the mixture have been physically separated, which is not the case.

The interferometer, and the absorption system with an electronic scanner, are of about equal convenience. The interferograms are in visible light, can be viewed directly and, with some experience, can be interpreted visually almost as easily as schlieren pictures. The absorption system using a television scanner (Lloyd and Esnouf 1974) is slightly more convenient than one using a photomultiplier, because the images of the cells can be seen directly on the monitor screen, and the plot of extinction versus the x coordinate is continuously displayed on an oscilloscope screen. With the photomultiplier systems the traces can be seen at any time by initiating a scan, when they then appear on the recorder paper.

The most inconvenient system must be the absorption system with photography using ultraviolet light. Before any indication of the results is available, the film must be developed and fixed, and by the time this is done it may be too late to alter any variables which had been inappropriately set.

Cost

The cost of installing an optical system in the first place must, of course, depend upon the quality of the components used. Since the schlieren system and the Rayleigh interferometer use exactly the same components the cost of having both is very little greater than having one, and manufacturers normally sell only a combined system. The inclusion of a cylindrical lens together with means to adjust the direction of its axis means that this system is likely to be slightly more expensive than the simple form of the absorption system. However, the more elaborate forms of the absorption system with monochromator and electronic scanner are extremely expensive and can easily account for 30 per cent of the toal cost of an ultracentrifuge. However, it must be remembered that the schlieren and interference systems will generally require ancillary measuring equipment, which should be included in the initial cost. Similarly, an absorption system using photography will need a microdensitometer for analysis of the photographs. Even so the cost of these pieces of ancillary equipment is unlikely to approach that of an electronic scanner.

The cost of operating the system is fairly small, amounting only to the cost of photographic materials and processing chemicals. In this context plates are more expensive than film but are more convenient when it comes to measurement. The electronic scanners save on the use of photographic materials but instead use recorder paper. The expense of this depends upon the type of recorder fitted and these may use special paper. This is particularly true of the television design of Lloyd and Esnouf (1974), which has a very fast recorder that uses a beam of ultraviolet light to write on photosensitive paper in order to handle the very rapid scanning speed of this design. The extra expense of this paper is to some extent offset by the ability of the system to show the output without using the recorder.

Conclusions

Undoubtedly the most common optical system in use today is the schlieren system. It is very simple to use and produces patterns almost ideal for the interpretation of migration experiments. For experiments of this type its comparative lack of precision on the z axis is of no consequence because it is only required to show the positions of the boundaries, and this it does with more than adequate accuracy. The effect of Wiener skewing is small and in most cases causes errors which are less than those caused by variations in other parameters. Since, for many systems, the effect of the concentration of the solute on its velocity of migration is small, the comparatively large concentrations required are no disadvantage. Indeed, in instruments other than the ultracentrifuge, smaller concentrations can often not be used because of the difficulty of stabilizing the boundary when there is only a very small change in the density of the solution across the boundary.

The recent increase in the popularity of the sedimentation-equilibrium technique for the measurement of molecular weights has resulted in an increase in the use of the interferometer in order to utilize the much greater accuracy of this system. The absorption system has only been used extensively in the past by those working on the sedimentation properties of the nucleic acids, whose large extinction coefficients at 260 nm enabled studies to be made at very low concentrations; this was important because the nucleic acids have a very large dependence of their sedimentation coefficient on concentration. The more recent development of the electronic scanners has removed many of the disadvantages of the simple form of the absorption system, and the multiplexers enable several different samples to be studied simultaneously.

These systems are new and have not yet come into very general use. However, the absorption system, coupled with an electronic scanner, is potentially the most versatile and powerful system at present available. There have also been advances recently in the use of computational techniques in the analysis of sedimentation experiments, and the ability of a scanner to be linked directly with a computer would enable these computations to be carried out directly, without the need for preliminary measurements of the photographs produced by the other systems. In addition the absorption system is potentially capable of studying some solutes at extremely low concentrations. This is particularly important to the biochemist because he is often concerned with the relationship between the physical state of a macromolecule and its biological function, which is usually manifested *in vivo* at extremely low concentrations. There is also an increasing awareness of the importance of the interactions of biological macromolecules with each other, and the ability of the absorption system to discriminate between solutes makes it most valuable in studies of these interactions.

REFERENCES

Akeley, D. F. and Gosting, L. J. (1953). *J. Am. chem. Soc.* **75**, 5685.

Antweiler, H. J. (1951). *Mikrochemie mikrochem Acta* **36/37**, 561.

Beams, J. W., Snidow, N., Robeson, A., and Dixon, H. M. (1954). *Rev. scient. Instrum.* **25**, 295.

Bethune, J. L. (1970). *Biochemistry* **9**, 2737.

Browne, C. A. and Zerban, F. W. (1941). *Physical and chemical methods of sugar analysis*, 3rd Edn, p. 1206. Wiley, New York.

Coulson, C. A., Cox, J. T., Ogston, A. G. and Philpot, J. St. L. (1947). *Proc. R. Soc.* A**192**, 382.

Dunlop, P. J. and Gosting, L. J. (1959). *J. phys. Chem.* **63**, 86.

—— and —— (1964). *J. phys. Chem.* **68**, 3874.

Ford, T. F. and Ford, F. F. (1964). *J. phys. Chem.* **68**, 2843, 2849.

Foucault, J. L. (1859). *Annls Obs. mun. Paris* **5**, 197.

Fujita, H. and Gosting, L. J. (1960). *J. phys. Chem.* **64**, 1256.

Gerhart, J. C. and Schachman, H. K. (1965). *Biochemistry* **4**, 1054.

Gosting, L. J. (1956). *Adv. Protein Chem.* **1**, 429.

—— and Akeley, D. F. (1952). *J. Am. chem. Soc.* **74**, 2058.

—— and Morris, M. S. (1949). *J. Am. chem. Soc.* **71**, 1998.

—— and Onsager, L. (1952). *J. Am. chem. Soc.* **74**, 6066.

Gouy, A. (1880). *C.r. hebd. Séanc. Acad. Sci., Paris* **90**, 307.

Gropper, L. (1964). *Analyt. Biochem.* **7**, 401.

Kegeles, G. and Gosting, L. J. (1947). *J. Am. chem. Soc.* **69**, 2516.

Kirschner, M. W. and Schachman, H. K. (1972). *Biochemistry* **10**, 1900.

Kruis, A. (1936). *Z. phys. Chem.* **34B**, 13.

Lamers, K., Putney, F., Steinberg, I. Z., and Schachman, H. K. (1963). *Arch. Biochem. Biophys.* **103**, 379.

Lamm, O. (1937). *Nova Acta R. Soc. Scient. upsal.* **10**, no. 6.

Lloyd, P. H. and Esnouf, M. P. (1974). *Analyt. Biochem.* In the press.

Longsworth, L. G. (1939). *J. Am. Chem. Soc.* **61**, 529.

—— (1947). *J. Am. Chem. Soc.* **69**, 2510.

—— (1959). In *Electrophoresis* (ed. M. Bier), p. 144. Academic Press, New York.

Lotmar, W. (1947). *Helv. chim. Acta* **32**, 1847.

Ogston, A. G. (1949). *Proc. R. Soc.* A**196**, 272.

Paul, C. H. and Yphantis, D. A. (1972). *Analyt. Biochem.* **48**, 588, 605.

Philpot, J. St. L. (1938). *Nature, Lond.* **141**, 283.

—— and Cook, G. H. (1947). *Research* **1**, 234.

Richards, E. G., Bell-Clark, J., Kirschner, M., Rosenthal, A., and Schachman, H. K. (1972). *Analyt. Biochem.* **46**, 295.

—— and Schachman, H. K. (1959). *J. phys. Chem., Ithaca* **63**, 1578.

——, Teller, D. C., Hoagland, V. D., Haschemeyer, R. H., and Schachman, H. K. (1971). *Analyt. Biochem.* **41**, 215.

—— and Schachman, H. K. (1971). *Analyt. Biochem.* **41**, 189.

Robkin, E., Meselson, M., and Vinograd, J. (1959). *J. Am. Chem. Soc.* **81**, 1305.

Schachman, H. K. (1957). *Ultracentrifugation in biochemistry*, p. 39. Academic Press, New York.

Spragg, S. P. and Goodman, R. F. (1969). *Ann. N.Y. Acad. Sci.* **164**, 294.

——, Travers, S., and Saxton, T. (1965). *Analyt. Biochem.* **12**, 259.

Svensson, H. (1939). *Kolloidzeitschrift* **87**, 181.

—— (1950*a*). *Acta chem. scand.* **4**, 399.

—— (1950*b*). *Acta chem. scand.* **4**, 1329.

—— (1951). *Acta chem. scand.* **5**, 1301.

—— (1954). *Optica Acta* **1**, 25.

——, Forsberg, R., and Lindstrom, L.-A. (1953). *Acta chem. scand.* **7**, 159.

Tanner, C. C. (1927). *Trans. Faraday Soc.* **23**, 75.

Thovert, M. J. (1902). *Annls Chim. Phys.* [7], **26**, 366.

—— (1914). *Annls Chim. Phys., 9ᵉ Serie Annls Phys.* **2**, 369.

Tiselius, A., Pedersen, K. O., and Erikson-Quensel, I.-B. (1937). *Nature, Lond.* **139**, 546.

Toepler, A. (1866). *Annln Phys. Chem.* ([v] 7), **127**, 556.

Trautman, R. and Burns, V. W. (1954). *Biochim. Biophys. Acta* **14**, 26.

Wiener, O. (1893). *Annln Phys. Chem. Neue Folge* **49**, 105.

Wiedemann, E. (1952). *Helv. chim. Acta* **35**, 2314.

Wollaston, W. H. (1800). *Phil. Trans.* **90**, 239.

Wolter, H. (1950). *Annln Phys.* [6], **7**, 182.

AUTHOR INDEX

SUBJECT INDEX

References to definitions are in *italic type*. Where more than one reference is given, the more important one is shown in **bold type**.